# The Chicken from Minsk

# The Chicken from Minsk

YURI B.
CHERNYAK

&

ROBERT M.
ROSE

TEXT ILLUSTRATIONS BY JOSEPH LATINSKY

BasicBooks
*A Division of HarperCollinsPublishers*

Published by BasicBooks, A Division of HarperCollins Publishers, Inc.


For information, address BasicBooks, 10 East 53rd Street, New York, NY 10022-5299.

Designed by Acme Art, Inc.

**Library of Congress Cataloging-in-Publication Data**

Chernyak, Yuri.
The chicken from Minsk / Yuri Chernyak, Robert M. Rose.
p. cm.
ISBN 0-465-07127-9
1. Mathematical physics—Problems, exercises, etc. I. Rose, Robert M., 1937- . II. Title.
QC20.82.C48 1995
530'.076—dc20 95-1874
CIP

95 96 97 98 RRD 9 8 7 6 5 4 3 2 1

## Contents

# Preface

"I've found the solution to this problem. I was thinking about it for two days, and I've finally solved it. I am so happy and excited. Listen, I am twenty-eight years old, I have a family and a child, and I've never been so happy in my life. I have wasted half my life. I could have been this happy when I was fourteen!"

This comment from an MIT student in the Concourse Program, which in 1992 began to use the kind of problems found in this book, is typical of the many we have heard. "If you sign up for this course," said another student, "do not make any other plans. Do not think you will have time to socialize, date, take showers, or sleep. If you do sleep, you will have nightmares about the problems—and they will haunt you in the shower too!"

Visiting the Concourse lounge early in the morning on the way to our offices, we often found our young problem solvers still at it from the day before. Known as the Five O'Clock Club for their tendency to stay up, exhausted but still arguing and analyzing, until that hour, these obsessed students gave us the idea to share these brainteasers with a wider audience.

The central problem is to teach the student how to think, rather than what to think. Many of our students have difficulty solving problems that require analytical thought, and many competitive measures of scientific ability resemble the board game of Trivial Pursuit, having little to do with scientific ability or the passions that lead to scientific or technical achievement. But solving problems, especially technical problems, is what the Massachusetts Institute of Technology is all about. Some five years ago a workshop was set up (by RMR) within the Concourse Program, dedicated to the solution of problem sets in chemistry and physics. The two of us met in the second year of this workshop, and began exploring the remarkable and ancient Russian universe of problems and problem sets. (We call it Russian because it existed long before the Soviets and was an integral part of the culture, a great recreation for those long winter nights.)

Under the Soviet rule that lasted for most of this century, state-sponsored contests requiring the solution of math and science problems became as popular as some sports are in the United States. The best high school students (typically, the top 10 percent) in the country participated in these contests, called Olympiads. Local winners advanced to higher-level competitions, and the very best among them competed for the national championships.

Even among the rank and file, it was not sufficient to learn the entire textbook and syllabus. This alone would have earned, at best, a grade of C. Students were expected to be able to use that material quickly and accurately as tools in a wide range of problems and applications, to have mastered the operational aspects: how to use what they knew.

The goal of mathematics education in particular was, in a very athletic sense, to train the brain. The need for athletic activities to train the body is recognized without question, but in some countries the need to train the brain is viewed, when it is viewed at all, with disdain. In Russia, because of the rigorous education, science and mathematics were enjoyable and even playful subjects. It would be no surprise to hear over lunch, "Do you remember that problem on the admissions examination eighteen years ago? I have a new solution!" For good scientists, doing science was filled with passion and joy.

Keep in mind also that in Russia scientists and mathematicians were national heroes, having streets, parks, and even cities named after them. This attitude was not restricted to Soviet scientists. A biography of Robert Wood, a noted American physicist, was a best-seller in Russia, available in any bookstore and reprinted at least 20 times. The best of the educational system in Russia was very much in the European tradition. The goal was not professional training but intellectual, moral, and ethical development. Euclidean geometry was taught from Grades 6 through 9, followed by one year of trigonometry, but never with any emphasis on practical applications. The purpose was to teach systematic thought and the perception of beauty in mathematics and other intellectual constructions. Also required *for everybody* was five years of physics, with the intention of conveying a general elementary understanding of the workings of the world.

The 100 problems in this book were selected primarily for recreation, but all of them will teach you something. Most of the science questions happen to involve Newtonian physics. In order not to obscure the central ideas, we have eliminated the need for calculus and even algebra whenever possible.

These questions may amuse you, puzzle you, stump you, or even cause intense and obsessive behavior. They were collected by Y.C. during his student days, his service as an associate professor of physics at Moscow State University, and his refusenik period (1976–89). Many are from university entrance exams and Olympiads, and from the very old Russian tradition of complex and deceptive riddles. From this huge pool (at least 50,000 questions), we selected the most interesting and relevant, and tested them on first-year students in the Concourse Program at MIT.

The warm-up chapters contain relatively easy problems intended to produce an appetite for the more demanding material that follows. In testing these problems in the Concourse Program, we emphasized formal analogies, symmetry, and invariance as major underlying ideas in physics, using familiar situations to introduce fundamental and important scientific concepts. As we did with our Concourse students, we have woven stories of scientific history and scientists throughout the book.

Briefly, we will explain here what the Concourse Program is. Founded 25 years ago, it is one of the many educational success stories at MIT. Concourse is based on the oldest idea in education: that a small community of scholars will find the best way for itself to learn. Every year, 64 matriculating students elect to enter the program, and all of their first-year requirements are satisfied by a tightly knit, highly competent teaching staff in a single classroom and a large lounge (with kitchen) for study, tutoring, socializing, and relaxing. This very congenial arrangement is located in Building 20, a dilapidated, comfortable, and beloved military-style building that had been thrown up in haste during World War II to house the developers of radar. It was in the supportive environment of Concourse that we explored the use of these problems with our students, on cold January mornings in a warm lounge, over breakfast in the kitchen.

No one should be considered fully educated without understanding how the world works. The beauty of mathematics and science must be demonstrated and understood. This book is a beginning. We hope you will enjoy it.

*Cambridge, Massachusetts*
*February 8, 1995*

# Acknowledgments

There are three voices in this book: that of Yuri Chernyak and, by extension, of all the unheralded Russian thinkers and students who first created these problems; that of Robert Rose, the director and shaper of the Concourse Program, which provided the motivation and environment for these problems to be developed in their present form; and that of the MIT students whose energy and enthusiasm encouraged us to proceed. Certainly, without the Concourse Program and its students, the experiment that led to this book would not have been conceivable. We also had the support and encouragement of the MIT Class of 1951, which were greatly appreciated.

We would like to thank individually the undergraduates who filled the Concourse lounge on those cold January mornings (and afternoons, and evenings!): the "tutors," Charu Chaudhry, Antony Donovan, Sushil Panta, and Mike Quirk; and the "students," Mike Allen, Becky Covert, Joshua Goldberg, Kwan Yong Ha, Charley Hamilton, Brad McKesson, Will Nielsen, Srivatsan Raghavan, Marion Shows, Van Van, Victor Washington, Daniel Weber, and David (Chia-Ying) Yang. All contributed, varying in degree and in kind, but all with generosity and enthusiasm, and for that help we are most grateful.

Much of the organization and editorial work for the manuscript was done by Van Van. The sketches in the text are the work of Joseph Latinsky. The sketch at the beginning of the book is by Lois Malone.

Many constructive suggestions were made by Dr. Maxim Umansky and Professor Jerome Y. Lettvin, for which we are grateful.

Finally, and most important, we acknowledge with gratitude the infinite patience of Natasha Chernyak and Martha Rose, to whom we happen to be (respectively) married.

# How to Use This Book

1. Try to solve the problem first *without* using the hints.

2. If you find any question, even a warm-up question, too difficult, go on to another one. Different people react very differently to the same question, and the reaction does not appear to be related to intelligence or education. Return to the difficult question later.

3. Read the solution only when you have arrived at a solution yourself, and even then only when you have thought it through completely.

4. The easiest questions are in the first two chapters and at the beginning of each subsequent chapter. The toughest ones are at the end. Again, you should realize that individual reactions vary widely. Do not be surprised if an "easy" question causes you difficulty.

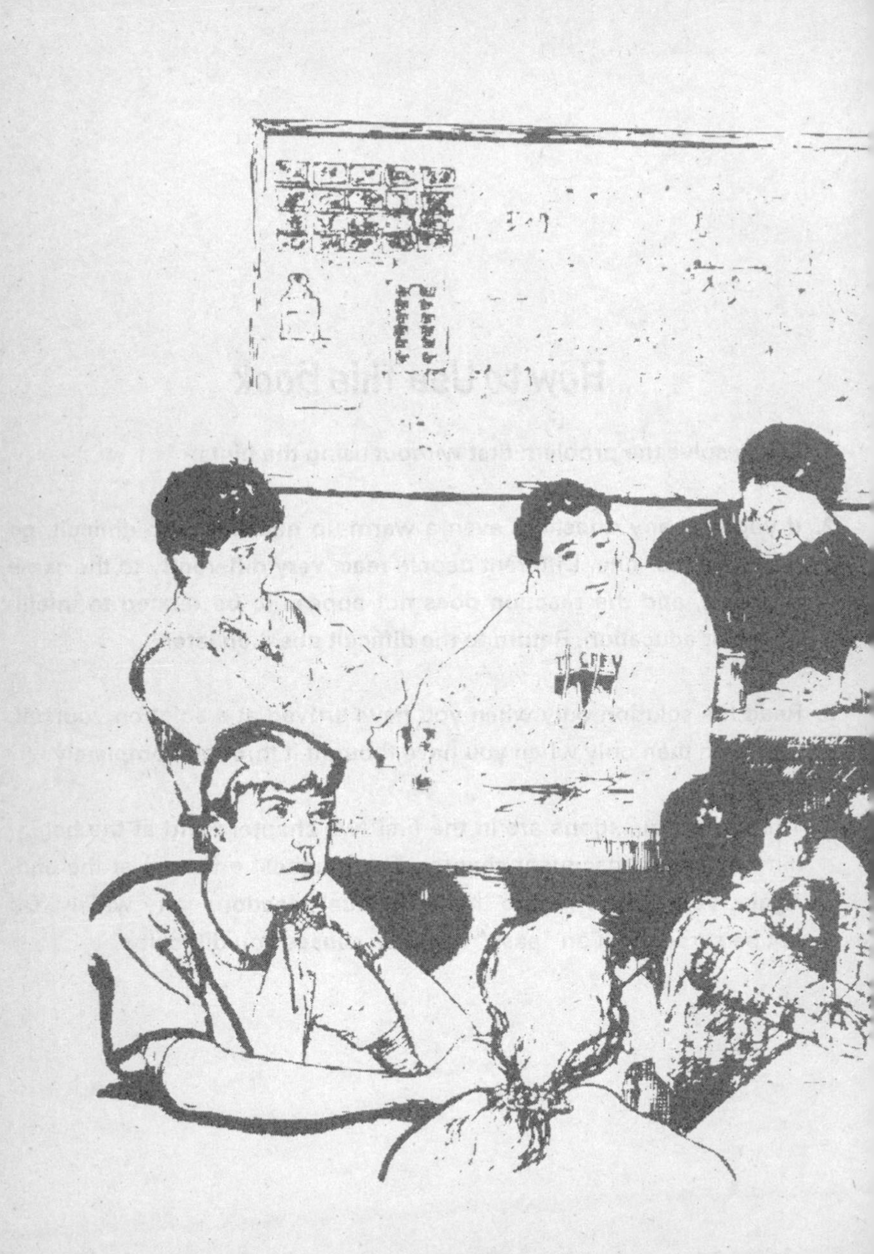

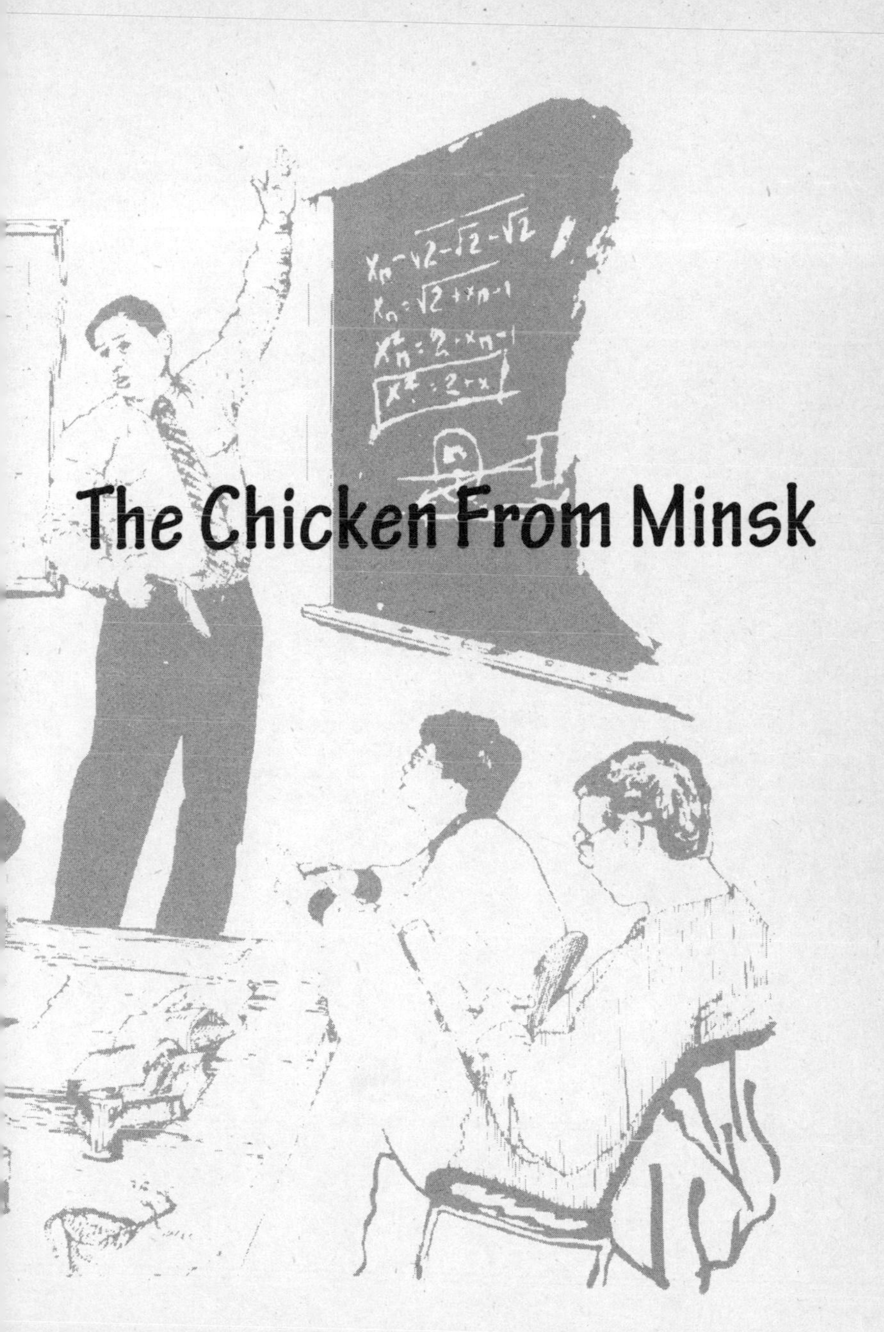

# The Chicken From Minsk

# Chapter 1

# Warming Up

## Natural Childbirth

1.

Two siblings are born naturally on the same date, in the same year, to the same mother and the same father. However, they are *not* twins—neither fraternal nor identical. Is this possible or impossible?

HINT:

*Pay careful attention to what is stated and what is not stated.*

# The Dumb Parrot (or The Problem with Mathematicians)

2.

The owner of a pet shop is a retired mathematician. He never lies, and he makes very precise statements. He tells a customer the parrot in the cage is extremely intelligent—in fact, "this bird will repeat every word he hears." The customer, impressed, buys the parrot. In a few days, the outraged customer returns with the parrot, saying, "I spoke to him for hours every day, but this stupid bird has not repeated a single word I said." Nevertheless, the pet shop proprietor did not lie. Is this possible?

**HINT:**

*There must be something wrong with the statement, the parrot, or the customer!*

## A Question of Art

**3.**

A sculptor named White, a violinist named Black, and an artist named Red meet in a café. One of the three says: "I have black hair, and you two have red hair and white hair, respectively, but none of us has a hair color that matches his name." White responds: "You are quite correct." What color is the artist's hair?

**HINT:**

*Try drawing a square table or matrix, with the names by hair color. Then try to fill it in in accordance with the problem statement.*

## To Eat or Sleep, That Is the Question!

**4.**

A problem in logic from the gulag: let us assume that one can survive *exactly* two weeks without food or without sleep. What should you do, eat or sleep, at the end of the fourteenth day without sleep and food?

**HINT:**

*It is impossible to eat and sleep simultaneously.*

## The Knights and the Pages

5.

Many years ago, three knights waited to cross the river Neva. Each knight had his own page, so there were six people. The boat they had could only carry two. However, the knights were ferocious killers and the pages were terrified. In fact, it was certain that any one of the pages would die of heart failure if he were not protected at every instant from the other knights by the presence of his own master. Was there any way to cross everyone over the river without losing a page?

**HINT:**

*Set up the situation graphically and analyze each move visually.*

## More Knights and Pages

6.

Suppose the previous problem involved four pairs of knights and pages. Is there a solution?

**HINT:**

*Prove that a situation similar to the fourth step in the solution of the previous problem (that is, with all the pages on one side of the river and the knights on the other) is inevitable. Then proceed from that point.*

# Yet More Knights and Pages: No Man Is an Island

7.

Reconsider the previous problem with four pairs of knights and pages, but with an island in the middle of the river. With the island, is there a solution?

**HINT:**

*By now you should not need hints to solve knight/page problems. In fact, you have now mastered medieval politics.*

# Grandfather's Breakfast

8.

Grandfather is a very hard-boiled customer. In fact, his eggs must be boiled for exactly 15 minutes, no more, no less. One day he asks you to prepare breakfast for him, and the only timepieces in the house are two hourglasses. The larger hourglass takes 11 minutes for all the sand to descend; the smaller, 7 minutes. What do you do? (Grandfather grows impatient!)

**HINT:** *Think carefully about the first step! The logic is unique!*

# The Prisoner and the Guards

9.

You are a prisoner with two guards, each guarding one door to your cell. One guard always lies and one is always truthful, but you do not know which is which. You may leave your cell by one of the two doors. One door leads to the execution block and death. The other leads to freedom. You may ask only one question, which you may address to either of the guards but not to both. What do you do?

**HINT:** *You must connect the guards and the doors with your question. (There is more than one solution to your dilemma.)*

# Ivanov and the Clock

**10.**

Ivanov is looking for his boss. Ivanov's boss is a Party functionary, and his office is luxurious, with soft furniture, thick carpeting on the floors, heavy draperies on the walls, and a thick, tightly closed, soundproof door. Usually, he is not in.

Ivanov opens the door to his boss's office and hears the wall clock strike once. The boss is not there. Ivanov leaves the door open. A half-hour later, Ivanov is astonished to hear the clock strike just once, again. After another half-hour, this experience is repeated; still, no boss. Now Ivanov, losing patience, waits another half-hour and hears the clock strike only once for the fourth time. The boss's secretary, who has now returned, tells Ivanov that the boss left for lunch shortly before Ivanov arrived for the first time. Ivanov knows that the clock strikes the number of hours on the hour, and strikes just once on the half-hour. When is the latest that the boss could have left his office?

HINT: Remember the carpeting, draperies, soft furniture, and soundproof door.

# Chapter 2

# More Warm-ups

## Shopping with Boris and Marina

**1.** Boris and Marina want to buy ice cream bars. However, Boris is 24 kopeks short of the price of a bar, and Marina is 2 kopeks short. They decide to pool their funds and buy a single bar. When they do, they find they still do not have enough money. How much does an ice cream bar cost?

**HINT:** *Just a comment here. The kopek is the minimum monetary unit in Russia, and the problem occurred often enough in reality there.*

## Shopping with Boris and Marina, Part II

2.

Boris and Marina are shopping again, this time for chocolate bars. Boris observes, "If I add half my money to yours, it will be enough to buy two chocolate bars." Marina naively asks, "If I add half my money to yours, how many can we buy?" Boris replies, "One chocolate bar." How much money did Boris have?

**HINT:** *If you have solved the ice cream bar problem, you may have certain suspicions regarding Boris.*

## At the Movies

3.

Olya, Petia, Kolia, and Klava regularly attend the movies on school days, usually on days when they wish to avoid a particular class or homework assignment. They began going together on one rainy day, and then went regularly but on different schedules. Olya went every fourth day; Petia every fifth day; Kolia every seventh day; and Klava every ninth day. When will they again all go together to the movies?

**HINT:** *The numbers 4, 5, 7, and 9 do not have common factors.*

# A Drinking Problem

4.

Besides chess playing and problem solving, drinking is and always has been the most common form of recreation in Russia. Vassily has acquired a 12-liter bucket of vodka and wishes to share it with Pyotr. However, all Pyotr has is an empty 8-liter bottle and an empty 5-liter bottle. How can the vodka be divided evenly?

**HINT:** *Just a remark here. Bottles were in short supply, particularly during the last stages of the Communist regime. Often, to fill a prescription it would be necessary to arrive at the pharmacy with an empty bottle.*

## The Caterpillar

5.

It is spring, and there are 12 hours of daylight. A caterpillar climbs a vertical wall at a speed of 1 foot per hour during the day, and then sleeps for the 12 hours of darkness, during which it slides downward at 1/2 foot per hour. The wall is 48 feet high. How long will it take the caterpillar to reach the top of the wall?

**HINT:**

*If the caterpillar reaches the top of the wall and naps, it will not slide down at all.*

## Washing Your Face

6.

It is possible to wash your hands with liquid nitrogen (77 Kelvin, or -196° C) without harm? Explain how this can occur.

**HINT:**

*If you experiment with a water droplet on a very hot frying pan, you will discover that the water droplet can be very long-lived.*

# Buying Berries

7.

Berries (many of which resemble cranberries) are popular in Russia. In one warehouse, 1,000 kg of fresh berries are stored. The berries contained 99% water when fresh, but a few days later, a test showed that there was now only 98% water, due to the drying out of the berries. What do the berries weigh now?

**HINT:**

*Make a rough estimate before you calculate. The problem involves an idea that is central to all of physics, that is, that your life (as a physicist!) will be much easier if you can identify a quantity that is conserved.*

# The Videotape

**8.** Rewinding an entire videotape cassette, where the tape spindle rotates at constant angular speed (that is, constant revolutions per unit time), takes 4 minutes. How long did it take to rewind the first 25% of the tape? For simplicity, assume that the diameter of the fully rewound tape reel is much greater than the diameter of the tape spindle.

**HINT:** *The length of the tape on the bobbin can be evaluated by dividing the surface area of the rewound tape on the side of the reel by the thickness of the tape. Also, the number of turns can be obtained by dividing the diameter of the rewound reel (less the bobbin diameter) by the thickness of the tape.*

## Adventures on the Moscow Subway

9.

Boris commutes to college by the Moscow subway, which runs in a circle. The school happens to be at the point on the circle that is exactly opposite to where Boris boards the train, so it takes the same time to get to school by train in either direction, and trains run in both clockwise and counterclockwise directions. The train schedules are very regular. The time interval between two successive counterclockwise trains is the same as for trains moving in the other direction; for instance, if there is an hour between the arrival of clockwise trains, there is also an hour between the arrival of counterclockwise trains. Boris observed, however, that he caught the clockwise trains more often than the counter-clockwise ones, despite the fact that his schedule was irregular and he arrived at the station at random times. Can you explain this?

**HINT:** *The train arrivals are not random. Boris's arrivals are.*

# The Crazy Dog (or The Problem That Did Not Fool John Von Neumann)

10.

Misha and Tisha are on their bicycles, a distance $L$ apart. They begin at the same time to move toward each other, each pedaling as fast as he can, intending to collide. At the instant they begin, their dog, who loves them both, leaves Misha and runs as fast as he can to Tisha, who pats him on the head. When this happens, he leaves Tisha and runs back to Misha, who also pats him on the head, at which point he turns and runs back to Tisha. All this is repeated until the bicyclists collide. How much distance is covered by this crazy, affectionate animal? Assume that Misha and Tisha move with constant speeds $v_1$ and $v_2$, respectively, and the dog moves with speed $u$ and is able to turn around instantaneously (he tries!).

HINT:
This is really a warm-up problem!

# Chapter 3

# Nikolai's Miserable Business Trip

## Cooling the Tea

**1.**

Nikolai travels to Vishny-Volochok on business. There are no kitchen facilities in the hotel there (in fact, there are no facilities of any kind), so Nikolai carries a small heating coil with him to heat his tea. Every morning he heats his tea, then puts sugar in it and waits for it to cool down to a drinkable temperature. On some mornings he is delayed, and his tea begins to cool before he puts the sugar in. Which way is faster?

**HINT:**

*Dissolution of sugar in water (or tea) is an endothermic reaction—that is, sugar absorbs heat when it dissolves.*

## Luxurious Accommodations

2.

Nikolai has to be in Saint Petersburg for a week. The accommodations are much better than at Vishny-Volochok, but also expensive. In fact, the ruble is not acceptable to the innkeeper. Nikolai has a silver chain with seven links, and the innkeeper agrees to accept silver at the rate of one link per day. He insists on payment for the day every morning, but also, to minimize damage to the silver links, that for the entire stay of one week, no more than one link can be cut. How can Nikolai pay the innkeeper?

**HINT:**

*Nikolai can give some pieces of the chain to the innkeeper and receive some others as change.*

## The Worst Case in Vishny-Volochok

3.

Back with Nikolai to Vishny-Volochok. The hotel clerk is, at the time Nikolai arrives with nine other travelers, quite drunk and has mixed up the room keys. There is no identification on any of the ten keys, and the ten rooms are locked. The ten travelers are very tired. What is the *maximum* number of trials (the worst case) required to sort out all the keys?

**HINT:**

*The number of trials required to sort out all the keys is not the same as the number of trials required to open all the doors. Consider how many trials you will need for the first key you choose.*

## Another Miserable Night in Vishny-Volochok

**4.**

Nikolai must leave the hotel very early in the morning to go to work. No room service, no restaurants. He and his friend make tea in their room (this is illegal, but they risk being thrown out of their room rather than go without breakfast). They both carry small immersion heaters that will heat a cup of water to boiling. Time is short, and it is necessary to make two cups of boiling water as quickly as possible. They can put their heaters in two separate cups and heat them simultaneously, or they can put both in one cup and heat two cups one after the other with the two heaters. Which method is faster?

**HINT:**

*If there were no heat loss, the time to heat the water would be the same for the two methods.*

# 5. Nikolai's Welcome Home

Nikolai finally returns home from Vishny-Volochok. His wife, Nina, notices deep straight gouges on Nikolai's brand-new luggage that she gave him recently as a birthday present. Angrily, she demands an explanantion. Nikolai responds with the following story. He was waiting for his luggage at the Moscow airport. His luggage, which has flat sides, was riding on a straight conveyor belt at constant velocity v. Vaska the cat escaped his owner and leaped directly across the conveyor belt, perpendicular to the motion of the luggage, but didn't quite make it. He landed on Nikolai's luggage, dug in his claws, and gouged the brand-new leather. Nina replies that

this cannot be true because the gouges are straight, and she is positive that a combination of the straight uniform motion of the luggage and the decelerating motion of the cat across it would produce a curved trajectory. Do you think Nina has caught Nikolai in a lie?

Consider two cases: one where the friction is negligible and Vaska slides across the luggage at constant velocity; and one where there is considerable friction (deep gouges in Nikolai's luggage) and Vaska decelerates, coming to a stop atop the luggage.

**HINT:**

*Your frame of reference is very important here! Imagine that you are standing on the luggage waiting for Vaska to jump. Also, you must ignore aerodynamic drag (this is another hint!).*

# Chapter 4

# Frames of Reference

## A Lesson in Aerial Combat

**1.**

In aerial combat, fighters generally attack bombers (and other fighters) from the rear. To protect military aircraft from attack from the rear, missiles were installed in the tail, pointing backward, to be launched toward any aircraft following behind. But when this arrangement was tested, in every case the missile made an immediate U-turn and hit the aircraft that launched it. (Luckily for the pilot, these test missiles had no explosive charges. However, the embarrassment was great.) Can you explain this phenomenon? (This is a true story. It occurred in the Soviet Union soon after World War II, early in the development of jet aircraft.)

**HINT:**

*At the time this device was developed, there were no electronic in-flight controls for such missiles; they were simple rockets, aimed and fired. Now consider the following: What is the kinematic difference between a shell fired from a cannon and a missile? Is the launching aircraft the proper body on which to fix the frame of reference for the missile?*

# Old Man Mazay Rows for His Vodka

2.

We are now with Old Man Mazay, a familiar figure in Russian folklore, rowing down a river. The current is 2 miles per hour. In the stern of the boat, in a wooden box, is the ever-present bottle of vodka. As he passes under a bridge he takes a drink, and then returns the bottle not to the box but to the river. Thirty minutes later he is ready for another drink and notices that the bottle is missing. Old Man Mazay rows (and consumes vodka) at a constant rate. In still water, on a lake, his speed is 3 miles per hour. How long will it take him to return to his still-floating bottle, and how far from the bridge will he be at that time?

HINT: This is another frame-of-reference problem!

# Fasten Your Seat Belt!

3.

One day not too long ago, I was flying on a small passenger plane from Boston to New York City. Since we were flying at an altitude of 10,000 feet, I had a wonderful view of the world beneath me, and my attention was held by a smaller, light plane that was crossing beneath us. The reason this plane caught my attention was that it traveled in a very awkward and surprising manner. I expected the propeller to pull the airplane straight ahead, in a direction parallel to the fuselage. Instead, it appeared to move at an angle to the fuselage, moving sideways as well as forward. The motion of the plane resembled, more than anything else, the way young dogs trot, at an angle to the direction in which they are facing. My neighbor in the next seat suggested that perhaps a crosswind was blowing the small plane sideways, but the pilot told us that there was no wind at that time. Can you explain what I saw?

**HINT:** *Strictly speaking, it doesn't depend on your point of view as much as on some other part of your situation.*

# Explosions

4.

A large number of balls is stuffed into a bag. In fact, there are too many balls. While suspended in the air the bag explodes, and the balls fly outward from the bag in all possible directions. Assume the balls all begin at the same point in space and all have the same speed, and ignore air resistance. What geometrical figure is described by the positions of the balls at any instant of time?

**HINT:**
*On the Fourth of July, after dark, look up.*

# Rainy Day on the Carousel

# 5.

The carousel you are riding on has no roof, and it begins to rain. You open your umbrella. What is the best way to hold the umbrella to keep you as dry as possible? You are at a distance $R$ from the center of rotation, the carousel rotates at constant angular velocity $\omega$, and the raindrops fall straight down with speed v (there is no wind).

**HINT:** *What is meant by "the best way to hold the umbrella"? To keep you as dry as possible, it must cut the biggest cross section out of the shower. Compare this problem with the soldier's problem in the May Day Parade* (problem #9).

## Misha and Tisha at the Carnival

6.

Misha and Tisha are also at the carnival. Misha's velocity is constantly changing but is *always* less than Tisha's velocity. Tisha's velocity also changes constantly but is always directed at the same angle to Misha's velocity. Also, Misha and Tisha are completely motionless relative to each other. Where are they?

**HINT:**
*The magnitudes of Misha's velocity and Tisha's velocity are both constant.*

## At the Carnival Again—Misha and Masha

7.

Misha and Tisha are still at the carnival. Tisha has wandered off in search of some food, and Misha is on the merry-go-round, riding a horse that is 2 meters from the center. There is another merry-go-round next to Misha's, with its center exactly 5 meters away from the center of the one Misha is riding, but revolving in the opposite direction at the same angular velocity $\omega$. On this second merry-go-round, also riding a horse that is 2 meters from the center of rotation, is Masha. Masha, a very attractive and highly competitive fellow student of Misha's at the Polytechnic Institute, finds that after passing him (in the opposite direction) several times she is

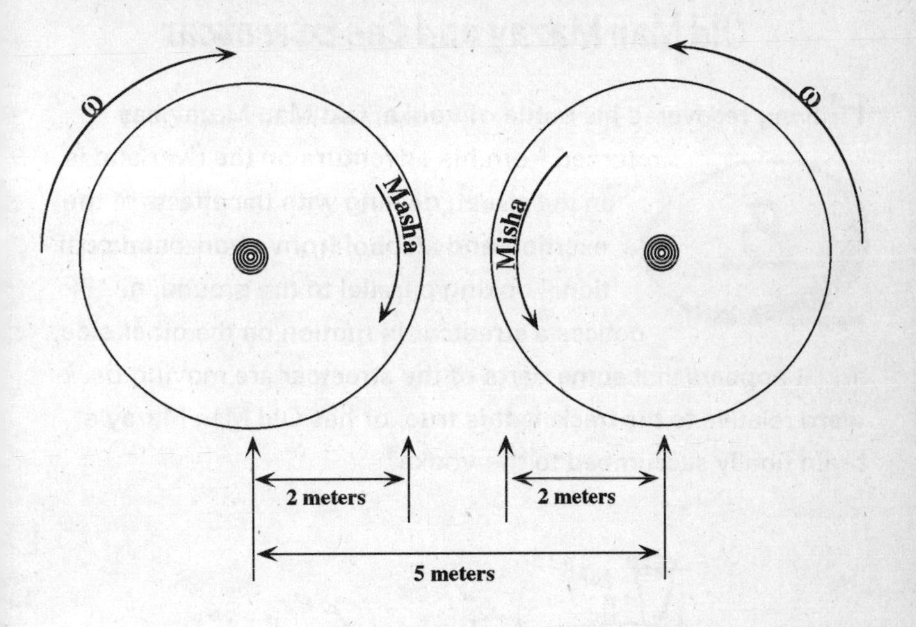


attracted to him, and they meet afterward. She remarks that his horse must have been slow, as she kept passing him. Misha replies that she must have been mistaken, as it was clear to him that his velocity was higher, and that *he* was passing *her* each time they approached. They were never able to resolve this issue, and what might have been a glorious romance was destroyed by their disagreement. Can you explain their observations?

**HINT:** *We are dealing here with noninertial, rotating frames of reference.*

# Old Man Mazay and the Streetcar

8.

Having recovered his bottle of vodka, Old Man Mazay has returned from his adventure on the river and is on the street, dealing with the effects of the exertion and alcohol from a horizontal position. Looking parallel to the ground, he notices a streetcar in motion on the other side, and it appears that some parts of the streetcar are moving *backward* relative to the track. Is this true, or has Old Man Mazay's brain finally succumbed to the vodka?

**HINT:** *This is definitely a frame-of-reference problem, in more than one respect. Consider particularly what parts of the streetcar Old Man Mazay is looking at. These railway wheels have flanges on the outside, which overlap the outside of the rails.*

# The May Day Parade

9.

You are a Red Army corporal preparing your mobile cannon for the May Day Parade in Red Square. There is no wind but it is, and will continue to be, raining. The cannon must proceed in the parade at constant velocity v. Your sergeant cautions you that the interior wall of the cannon barrel *must* not get wet at any time. For the sake of appearances, the Politburo will not tolerate any cap or plug in the barrel; the public must be permitted to stare down the gaping barrel of your weapon. Also, you must not drape anything over the outside of the cannon, for the same reason. Not wishing to spend any time in the gulag, you assure your sergeant that you will keep the interior wall of the cannon barrel dry without any covering or plugging. How do you do it?

HINT:
*Look at the problem from the point of view of the cannon.*

# With a Roll of Drums, the Galilean Frame of Reference

10.

A cylindrical drum with radius $R$ rolls between two parallel plates without slipping. The upper plate moves to the right with velocity $\mathbf{v}_U$ and the lower plate moves to the left with velocity $\mathbf{v}_L$, with both velocities parallel to the planes of the plates. If $\mathbf{v}_U = -\mathbf{v}_L$ and $|\mathbf{v}_U| = |\mathbf{v}_L| = v$, how many revolutions per second does the drum make? Also, how many revolutions per second are made in the general case $|\mathbf{v}_U| \neq |\mathbf{v}_L|$ ?

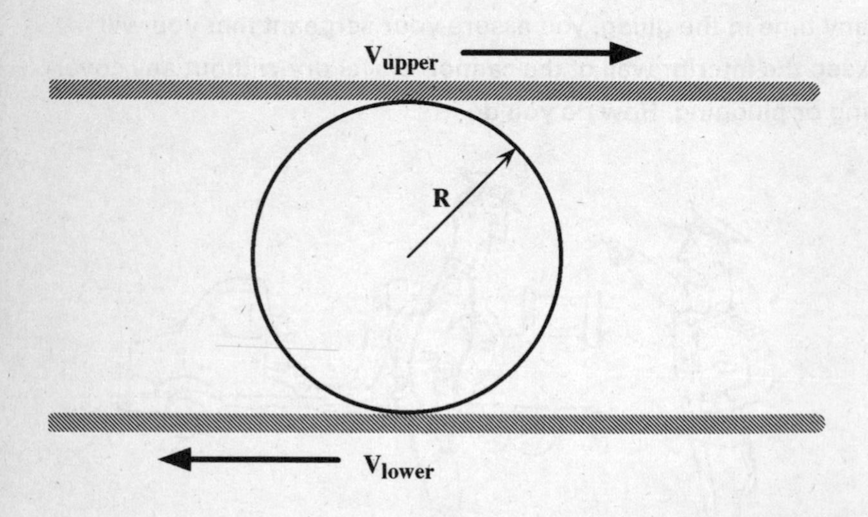


**HINT:** *You can choose the frame of reference so that the motion of the plates would look symmetrical. However, a better way is to use the lower plate for the reference frame.*

# Boris and the Wet Basketball—Reference Frames and Fluxes

**11.**

As Boris is playing basketball it begins to rain, and he reluctantly returns home, sliding the basketball ahead of him. His friend Tisha calls out to him to move faster, telling him that the ball (and Boris) will not get as wet if he gets home faster. Boris refuses, saying that the faster the ball moves, the faster it will get wet. There is no wind, and the rain falls straight down with a speed of $u$. The ball is sliding along horizontally with a speed of v. Who is correct?

**HINT:**

*Boris and Tisha are not making the same statement. Begin by thinking only of the total rate at which raindrops hit the ball, rather than how wet Boris and his ball will be when they get home. Then reexamine the question. Is the moving ball hit by a greater number of raindrops per unit time than a ball at rest?*

## Chicken Feed

12.

Boris is pulling a wheelbarrow with a bucket of chicken feed in it. It begins raining. Boris begins to run, to avoid getting the chicken feed wet. Does this help? Does the bucket fill with rain faster when Boris walks or when he runs? Assume that the bucket remains upright with its opening horizontal, and that the rain falls uniformly and at constant (terminal) velocity. There is no wind.

**HINT:**

*From the interior of the bucket, look up.*

## Fluxes and Conservation Laws (or It Always Helps to Run in the Rain)

13.

In the previous two problems you have seen how Boris's basketball would get wetter at a greater rate if he were to move faster, but his chicken feed would get wet at the same rate. Is there an inconsistency? If not, can you explain it?

**HINT:**

*It is not due to a fundamental difference between basketballs and chicken feed, but rather to an important property of the flux.*

# Chapter 5

# On the Road

## The Cossack and the Goat

**1.**

An important element in the training of Cossacks is the use of the lariat, or lasso, to trap animals (or people!).

However, because the training is under the supervision of the Party, there are certain restrictions, as shown in the diagram on page 36.

The Cossack is ordered by the Party Secretary to stay on one road, which is perfectly straight. The Cossack's horse proceeds at a constant speed down the road (the horse is a Party

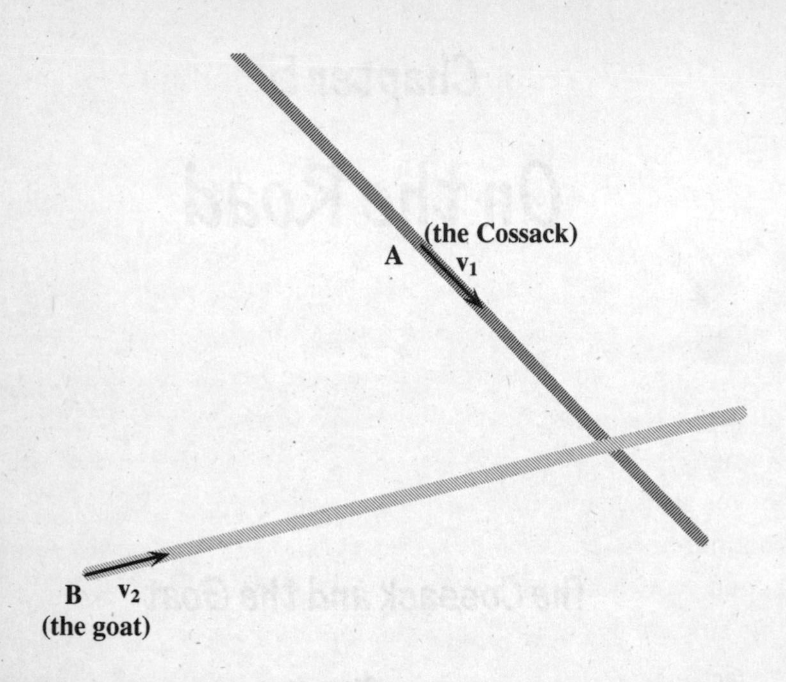


member). Now the Cossack sees a stray goat proceeding down an intersecting road, also straight. Because the goat used to belong to the Party Secretary, it also moves with constant velocity. How long must the Cossack's lariat be? That is, find the minimum distance between the Cossack and the goat, assuming the Cossack starts at point A and has velocity $v_1$ and the goat starts at point B and has velocity $v_2$.

**HINT:**

*The most baffling thing about this problem is usually a misconception of what is really given, and what formulation is needed for a solution. If you believe that the motion is fully given and the figure is adequate, you can proceed graphically, with data taken directly from the figure. You can also find an analytical solution and consider all the necessary parameters in the form that is most convenient to you. In fact, if you have had any experience in the operation of radar in aircraft or boats, this problem will look very familiar (and obvious!) to you.*

## A Triumph of Soviet Planning

2.

Every 10 minutes a truck filled with coal leaves Kemerovo and moves with constant speed (50 miles per hour) over a perfectly straight road to Magnitogorsk, which is 50 miles away.* Because the last 5-year plan requires that some coal be delivered to Kemerovo as well, a single truck loaded with coal also leaves Magnitogorsk for Kemerovo, traveling in the opposite direction. This single truck leaves at the same time as one of the Kemerovo-to-Magnitogorsk trucks and travels on the same road at the same speed. How many of the Kemerovo-to-Magnitogorsk trucks will this single truck pass during its trip?

***In fact, the distance is much greater. We have understated the distance to make the problem look simpler (it isn't!).**

**HINT:** *Examine our statements with care.*

## Backing into the Answer

3.

If you drive, you know from experience that it is far easier to parallel-park a car by backing in rather than by pulling in by forward motion. Explain why this is so.

**HINT:** *Suppose you were pulling out of a parallel parking space. Is it easier to back out or to move forward? Now play this maneuver backward, on an imaginary videotape.*

## Another Driving Problem

4.

Referring to the previous problem, you may have also noticed, from your driving experience, that no matter how carefully you drive, the front tires wear out faster than the rear tires. Explain why this is so.

**HINT:** *Consider your answer to the previous problem. In particular, think about the rotation of the front tires while the car is being steered.*

## How Far Did That Car Go?
## (A Trip Through the Wrong Dimension)

5.

A race car is tested on a straight track. From a standing start, it is accelerated to its maximum velocity. The brakes are then applied and it is decelerated to a stop. Because the car was designed by mathematicians, its performance is highly symmetric. In fact, when the velocity is plotted versus time, the plot turns out to be a semicircle, as shown in the figure on the facing page.

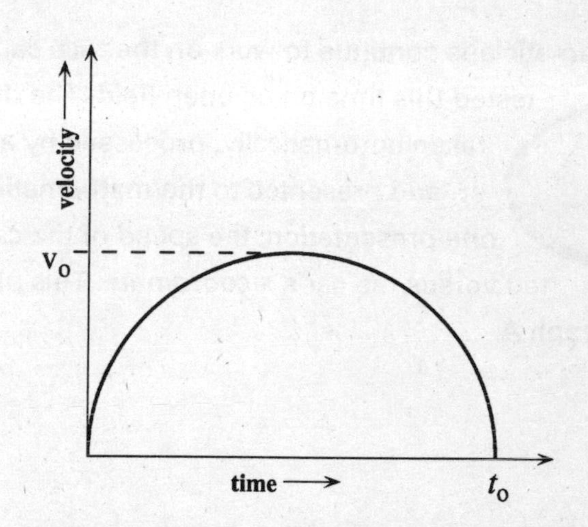


The maximum speed is $v_o$ and the total time, from go to stop, is $t_o$. Now it is necessary to calculate how far the car traveled. All the mathematicians conclude that the distance traveled is just the area under the velocity versus distance curve, or in this case just the area under the semicircle. But half the mathematicians, using the usual formula for the area of a circle, conclude that the distance traveled must be $(\pi/2)(v_o)^2$, since $v_o$ is the radius of the semicircle. The other half observe that $t_o/2$ is also the radius of the semicircle, and therefore the distance traveled must be $(\pi/2)(t_o/2)^2$. The atmosphere turns ugly. The custodian, on overhearing the animated discussion, observes that both groups are obviously incorrect. Who's right?

**HINT:**
*The ordinate and the abscissa of this plot have different dimensions!*

# Where Did That Car Go Now?

## 6.

Our mathematicians continue to work on the race car, which is tested this time on an open field. The data are taken automatically, processed by a computer, and presented to the mathematicians. In one presentation, the speed of the car is plotted versus the car's *x*-coordinate. This plot is shown in graph A.

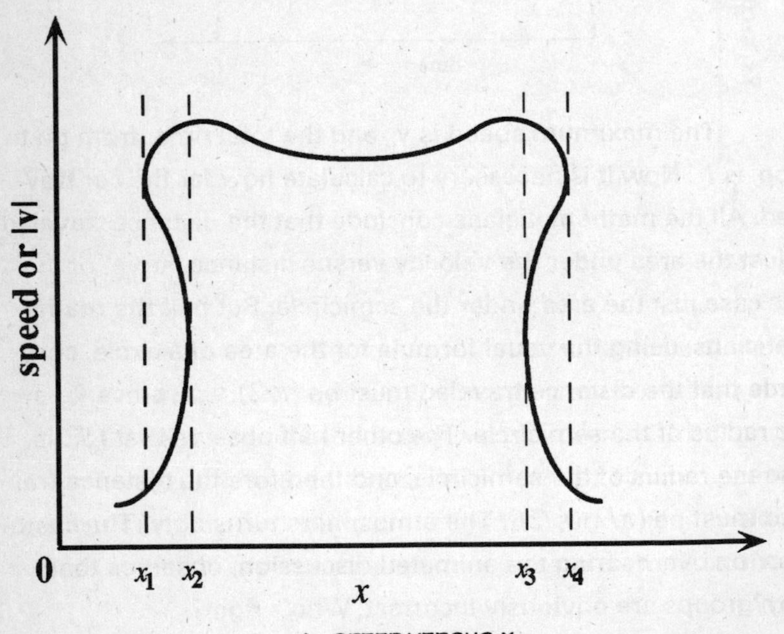


A. SPEED VERSUS X

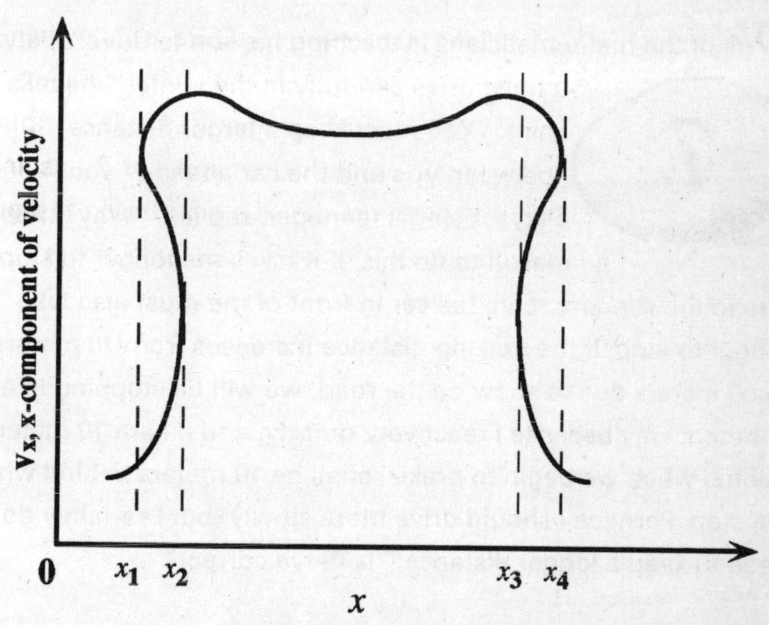


B. X - COMPONENT OF VELOCITY VERSUS X

In another plot the *x*-component $v_x$ of the velocity of the car is presented versus the car's *x*-coordinate. This plot is shown in graph B. The custodian, who has been apprehensive about these tests since the previous discussion (see the preceding problem), points out that there must be a mix-up in the software, as one of the two graphs is obviously incorrect. Which one, and why?

HINT:

*Think about the variables presented and how they must behave.*

## Winter Driving

7.

One of the mathematicians is teaching his son to drive. "Petya, you must drive carefully in the winter," he tells him. "You must keep a larger distance between you and the car ahead of you." Petya, being a teenager, replies, "Why? I see no reason to do this. If it takes me longer to stop due to the slippery road, the car in front of me must also take longer to stop. If the braking distance increases from 10 meters to 60 meters due to snow on the road, we will be stopping in a coherent way because I react very quickly, and if I am 10 meters behind when we begin to brake I shall be 10 meters behind when we stop. Perhaps I should drive more slowly, but I certainly don't need to keep a longer distance." Is Petya correct?

**HINT:** *Consider the conventional wisdom carefully, as it is much safer to follow at longer distances in winter. Petya is correct only under certain assumptions that do not hold in winter.*

## Winter Driving II

**8.**

Is it really necessary to drive more slowly in winter to avoid skidding, or is this advice simply meant to give you more time to react?

**HINT:** *Consider carefully the mechanics of skidding, particularly the friction force with the road, which cannot exceed a maximum value.*

## Another Triumph of Central Planning

**9.**

Vodka is produced in large amounts in Chekhov, and consumed in Serpukhov. It is shipped in large vertical cylindrical tanks (actually slightly washed-out oil tanks) with radii of 11 meters, which are permanently mounted in the upright position on the beds of trucks. These are so large that the trucks cannot pass each other on the single narrow, winding road between the towns, so a new road is planned. The idea is to use one road to ship vodka from Chekhov to Serpokhov, and the

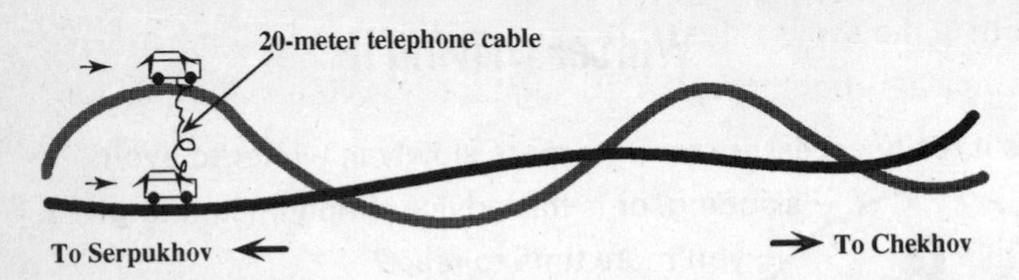


The bureaucrats travel from Serpukhov to Chekhov, connected by a 20-meter telephone cable

The vodka goes from Chekhov to Serpukhov on trucks with vertical cylindrical tanks with 11-meter radii. Empty tanks return on the other road.

other for the empty tank trucks to return for more, as shown in the lower portion of the figure.

The new road is also narrow and winding, but to allow for easy inspection by suspicious bureaucrats, the new road is planned in the following way. Two cars full of bureaucrats depart from Serpukhov simultaneously, as shown on the upper portion of the figure. One car travels on the old road and one on the exact route of the new road, connected by a telephone line 20 meters long (wireless communication was not available at the time, and in any case not secretive enough). Thus the centers of the cars cannot be further apart than 20 meters (they can be closer), although the roads themselves can twist and turn and even cross each other. No overpasses are possible. The two roads are both at the same altitude. The engineer in charge of the project complains about this plan, stating that it will be impossible for the trucks to pass each other and that the new road is totally useless, as only one road can be used at a time with the trucks. The entire city of Serpukhov, faced with an interruption of its vodka supply, is in danger of dying of thirst.

The Commissar in the Central Planning Bureau, citing the necessity for efficient surveillance, says that the trucks need not travel as the official cars did, they will not have any telephone line between them, and they can wait and avoid each other by staggering trips and by suitable maneuvering along the winding roads. Certainly it is clear that, because the bureaucrats' cars were connected by 20-meter telephone wires, at each point on one road there is a corresponding point on the other road that is no farther than 20 meters away. When a truck is at such a point, the bureaucrat argues, it is a simple matter for the other truck to be elsewhere, and not at the corresponding point that is no more than 20 meters away. In any case, he says, he will not accept the engineer's statement unless he can rigorously prove that it is absolutely impossible for two trucks to leave simultaneously, one from Serpokhov going to Chekhov and the other from Chekhov to Serpokhov, and reach their destinations without collision. Can you prove it?

**HINT:**

*Since the Commissar is not willing to accept your intuitive answer, you must present a rigorous proof, because much is at stake. Because you are not interested in the velocities of the cars or the trucks, you can completely define the instantaneous situation if you specify two distances along the two roads.*

# Chapter 6

# Sources and Sinks

## A Plumbing Problem

1\. A faucet has been left slightly open for some time, and a gentle stream of water flows downward. Why does the stream become thinner as it gets farther away from the faucet?

HINT:

*The stream has been flowing steadily for some time and does not change its shape.*

# More Practical Advice from Your Plumber (How to Measure Gravity with a Ruler and a Bucket)

2.

Consider the stream of water in the previous problem, becoming thinner as it emerges from the tap. Can you find the volumetric flow rate (e.g., gallons per minute or liters per second) and the velocity of the stream by using only a ruler? (If you insist on greater precision, use a caliper.) Can you use this technique to measure the acceleration of gravity $g$?

HINT:

*You will have to use conservation of energy here, as well as conservation of mass.*

## "You Have 20 Seconds to Make Your Calculation. Then I Strike!"

3.

Before it attacks, a cobra usually rises so that the head and the body following the head are vertical; the tail remains horizontal and at rest as the head rises. Find the force exerted on the ground by the cobra's tail. Assume the cobra has mass $M$ distributed uniformly along its length (like a garden hose) $L$, and that the head rises vertically with constant speed v.

HINT: *Consider Newton's Second Law in its most basic form, which does not necessarily involve an acceleration. The Second Law balances the rate of change of momentum with the vector sum of the forces on the body.*

## Conserving Momentum (and Energy!) in St. Petersburg

4.

Once upon a time there lived, in St. Petersburg, two street sweepers. One sweeper, Oblomovetz, was lazy, and the other, Stakhanovetz, was a workaholic, but in all other respects they were identical.*

***Stakhanov was a hero of the Stalin era because of his reputed enormous productivity, energy, and capacity for work. Oblomov, as described in a novel by Ivan Goncharov, needed an entire day just to get out of bed.**

One particularly bad winter day, all the electricity in town failed and snow began to fall. In order to get to work, the sweepers were launched on identical frictionless trolleys that left the station at the same speed ($v_0$). (The frictionless trolley is an achievement of the Party and modern socialism.) Oblomovetz immediately went to sleep. Stakhanovetz immediately began to sweep the trolley clean, throwing off the snow as soon as it landed, perpendicular to the direction of his motion.

Which of the two street sweepers traveled the farthest distance in the same time? There are no hills in St. Petersburg; the landscape of the city is perfectly flat. The snow fell at a constant rate of $\mu$ kilograms per second on each trolley.

HINT:
*Consider the momentum of each trolley.*

# Mikhail's Merry-Go-Round

5.

Mikhail manages the carousel, and he notices that during periods of rain, his electric bills are higher. Suppose the rain falls straight down, with no wind, the speed of the raindrops is $v$, and the density of the rain, in mass of water per unit volume, is $\rho$. How much additional power is needed to keep the carousel rotating at constant angular velocity $\omega$? Assume the rainfall to be uniform and the radius of the carousel to be $R$.

HINT:

*How does the rainwater leave the carousel?*

# Catching the Bus to Yakutsk (While Finding the Mach Cone)

6.

The bus to Yakutsk runs with constant velocity v over a perfectly straight road through a perfectly flat field (if you consult any reference on the region, you will see that this is another Socialist dream). It is cold (Yakutsk is in Siberia), and the instant you see the bus (say, at time $t = 0$) you will start running toward the road in order to board the bus. Your maximum running speed is $u$.

(a) Describe the region you should be in, in order to catch the bus. You may represent the road by a horizontal line, and if you wish, assume $v/u = 2$ for a sketch of this region.

(b) Imagine a large number of identical people (still another Socialist dream) uniformly spaced along the boundary of the region you have described, which we will call the "catching-up region." Now imagine you have a moving picture or videotape of these people running to the bus and boarding it (the bus can accommodate an infinite number of people). Finally, run this movie *backward*. What will you see? Do you know of any natural phenomena that resemble this picture?

**HINT:**

*The question is helpful enough. Hurry up, or you'll miss the bus!*

# An Explosive Topic: The Ice Cream (Mach) Cone

7.

A thin rod with semi-infinite length is made of a high explosive. The detonation speed—i.e., the velocity with which the detonation propagates down the rod—is v. The speed of the explosion products through space is $u$, and $u$ is less than v. Describe the shape of the explosion, in space and in time. That is, what is the time evolution of the region occupied by the explosion products?

**HINT:** *A global solution will give you the big picture.*

# More on High Explosives (and Ice Cream)

8.

Consider the previous question and its answer. What will be the shape of the explosion if the rod has finite length $L$? What will be the shape of the explosion if the propagation speed v in the finite rod is *less* than the speed $u$ of the explosion products in space?

**HINT:** *Use the same approach as the solution to the previous problem.*

# Everything Is in Flux (The Quick and Dirty Helicopter)

9.

A model of a helicopter at your design bureau can perform satisfactorily with a 50-watt motor. The real helicopter will be scaled up from the model by a factor of 10. A fellow engineer very quickly estimates that the power required for the real helicopter will be approximately 160 kW (for those of you who still cling to the British system, about 210 horsepower). How did he arrive at his quick, dirty estimate?

**HINT:** *For the scale-up, the surface area of any part of the helicopter should be proportional to the square of the length, and the mass should be proportional to the cube of the length. Also consider Newton's Second Law in its most general form, in terms of momentum.*

# Body Heat

## 10.

Boris has just learned thermodynamics and he is also, as you may have gathered from some of the warm-up problems, preoccupied with money. He is paying the electric bill for his apartment when Marina enters. He says: "It is very expensive to increase the energy of this apartment in the wintertime. I don't know what happens to all the energy that goes into the apartment, and I don't know why I am paying all this money. After all, there is very little in this apartment besides me!" Marina (who is a dancer and has had no technical education) replies: "It is very simple. You are paying to keep the apartment warm. It has nothing to do with energy." Do you agree with Marina? Where does the energy go?

**HINT:** *Pay careful attention to how energy depends on temperature, and to the contents of Boris's apartment.*

# Chapter 7

# Expanding and Contracting Universes

## Holding the Line on the Ruble

1.

Boris has drawn a straight line on a coin. The authorities have assured him that the coin is made of a homogeneous material. Boris is convinced that if the coin is heated up, the line will become curved, as shown in the figure, because some parts of the line are farther from the center of the coin and therefore will be pushed farther out. Marina disagrees. Who is correct? Can you prove it? *(See next page.)*

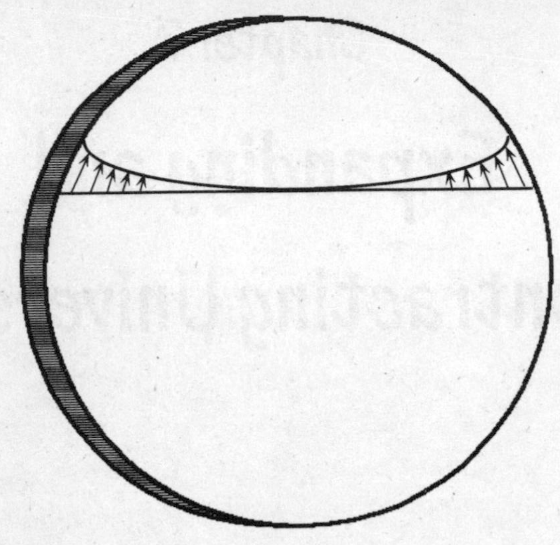

**HINT:**

*The coefficient of linear thermal expansion α is defined by the relation $l = l_0 (1 + \alpha \Delta T)$, where $\Delta T$ is the change in temperature and $l$ is the length of an object or the distance between any two points on the object.*

## Inflation of the Currency

**2.**

Boris has a copper coin with a hole drilled in the center. He has learned about thermal expansion and tells Marina that if the coin is heated, the metal will expand in all directions, inward as well as outward, so that the hole must get smaller as the temperature rises. Marina expresses her doubts. Who is correct?

**HINT:**

*Marina once worked in a bakery.*

# Expanding Bureaucracy

3.

High in the Kremlin, Brezhnachev and Chernopov are having an intense debate on energy conservation: Should lamps and other electrical appliances rest on tables or other supports, or should they be suspended? Which approach will consume less energy? If the latter is true, then the Party will require, for example, all lighting fixtures to be hung from the ceiling; table lamps and floor lamps will not be permitted. Here is the essential question: consider two identical spheres of any material, one suspended by a fine, insulating thread from the ceiling, and the other resting on a flat, insulating table. Which will require more energy to heat it up? You may ignore any conduction of heat to the thread and table. Consider both the thread and the table to be totally inert and motionless, not participating in the process in any way except to suspend or support the spheres.

HINT: *We have given you enough hints already!*

# Old Man Mazay Welcomes You to the Friedman-DeSitter Universe! (Is the Hubble Constant Really Constant?)

4.

Old Man Mazay, perhaps the first ecologist described in literature, was the main character in a popular Russian short story of the 1800s. Each spring, when the snow from the previous winter melted and flooded the fields, Mazay would go around rescuing hares from tiny islands of dry land. Whenever Mazay brought a group of hares to safety and released them, they would quickly scamper away in different directions. Being Russian hares, they behaved in a mathematically precise way, as shown in the figure.

Each of six hares run with the same speed v and, in order to spread out, at 60° angles to the hares on either side. As they run, the most intelligent hare looks back and is greatly sur-

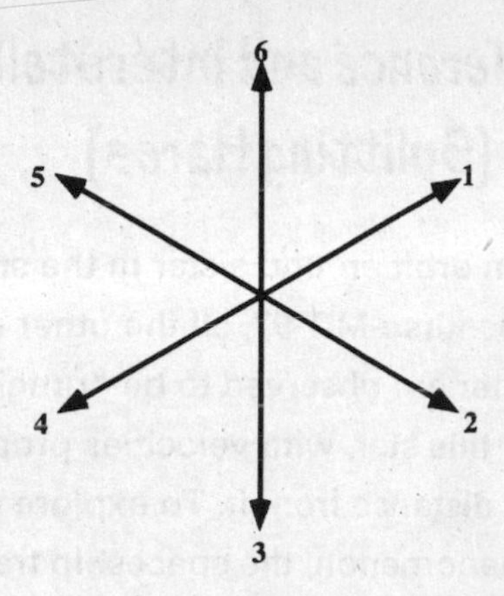


prised by what he finds. We ask that you determine what he found—in particular: How fast were the other hares moving (relative to him), and was there any relation between their velocities and their positions (again, relative to the most intelligent hare)?

HINT:

*Begin with a Galilean transformation in order to look at the relative velocities.*

# Frames of Reference and Interstellar Motion (Splitting Hares)

**5.**

From a spaceship in orbit around a star in the stellar cluster Concourse-MIT-97, all the other stars in the cluster are observed to be "running away" from this star, with velocities proportional to their distance from it. To explore this remarkable phenomenon, the spaceship travels to one of the other stars in the cluster, a star with a high velocity relative to the original. What do the astronauts observe from this star?

**HINT:**

*Distances and orbital movements within any solar system are negligible compared to interstellar distances. The rest of the answer lies with Old Man Mazay's hares.*

# Thermal Expansion and Cosmological Cats

6.

Three cats sit on a large sheet-metal roof as it warms up in the morning sun. They watch each other very carefully, as cats do. Cats excel at noticing even small movements, and these cats notice that they are moving away from each other as the sheet metal warms up—in fact, each cat moves away from each of the other cats at a rate that is proportional to the distance between them. The two cats who are farthest away are moving the fastest away from each other; the two closest cats separate at the slowest rate. Show that this is a consequence of the linear coefficient of thermal expansion of the metal. Assume that the sheet metal is heating up at a constant rate. The linear thermal expansion coefficient $\alpha$ is defined by the equation $l = l_o(1 + \alpha \Delta T)$ between any two points, where $\Delta T$ is the temperature change and $l$ is the length of an object or the distance between any two points on the object.

HINT:

*Examine carefully the definition of the coefficient of linear thermal expansion.*

# Masha's Mathematical Turtles

**7.**

Masha has trained her four turtles to always follow each other. She arranges them at the corners of a square, as shown in the figure, with each turtle facing its clockwise neighbor. The turtles move at only one constant speed v. What happens to the square? When do the turtles meet?

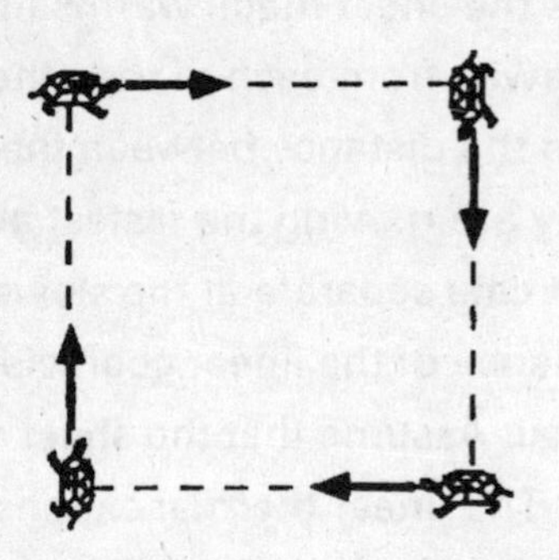

**HINT:** *Symmetry is very important here. Due to the symmetry of the arrangement, the turtles will always be arranged at the corners of a square.*

# Chapter 8

# Learning the Ropes

## Aleksandr and the Commandant

1.

Aleksandr is the skipper of the fishing boat Olga. He pulls into the dock at Tikhvin with his holds full of fish. Kretinov, the Commandant of the port, demands to know how many kilograms of fish are in the boat, for reporting to the Party Secretary. However, the facilities in the port for weighing boats and for unloading fish are not in working order. Aleksandr proceeds to measure the weight of the boat with a fish scale! The scale has a capacity of only 100 kg, and the boat is very large—perhaps 1,000 times or more the capacity of the scale. How does Aleksandr accomplish this measurement? Aleksandr's engineer slyly remarks to the first mate that Aleksandr has swindled Kretinov again, that the weight has been overstated. Do you agree?

**HINT:**

*Isaac Newton would have been pleased with Aleksandr's method, as it uses his Second Law very well. However, the results are somewhat watered down.*

# Unloading Aleksandr's Fish

2.

After Aleksandr has convinced Kretinov that he has correctly measured the weight of his fish-laden boat, he proceeds to supervise the unloading. Two fixed pulleys at the same height are used to lessen the strain on the ropes, as shown in the figure. The crate of fish is placed at a location equally distant from each pulley, and the ropes, which are of exactly the same length, are pulled at exactly the same speed v. This should divide the load equally between the two ropes and the two pulleys. Kretinov has recently learned about such situations (having taken a course entitled "Physics for Bureaucrats") and understands vector addition. He concludes that the velocity of the fish is a vector sum of the velocities of the ropes, so that the crate of fish should rise with the speed $u = 2v\cos\alpha$, where $\alpha$ is the angle between either rope and the vertical line drawn along the path of ascent of the fish. Is Kretinov correct, or is there something fishy about his calculation?

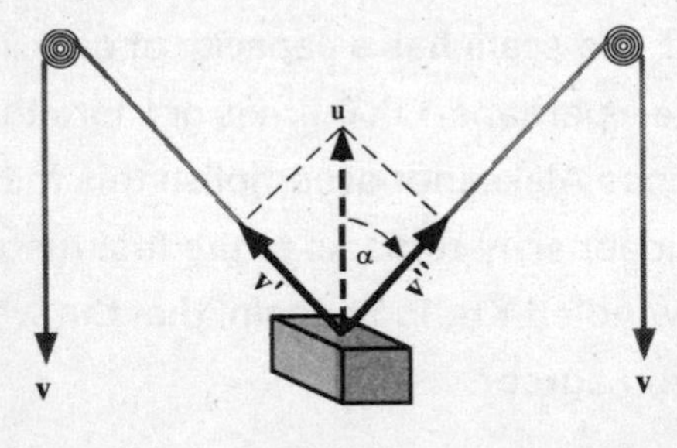


$|\mathbf{v}'| = |\mathbf{v}''| = v$

**HINT:** *Will this problem be solved by the resolution and addition of vectors? Examine the figure very carefully!*

## Aleksandr's Fish Go Supersonic

3.

Examine the solution to the previous problem. In particular, look at the velocity of the fish crate when the angle $\alpha$ approaches 90°. In fact, as $\alpha$ approaches 90° the speed should become very large, and it appears that Aleksandr's fish may be able to do what Sasha could not (see chapter 10, problem #2) — that is, exceed the speed of light. Is this possible?

HINT:

*If you have examined the solution to the previous problem, you may want to call this "Kretinov's revenge."*

# Stuck in the Mud (The Politically Correct Rope)

You are driving on an unpaved road when your truck bogs down in the mud. Russian soil is either clay or *chernozem,* a very rich, organic, fatty mixture. When wet, either soil is mechanically unstable and very slippery. Because dirt roads are the rule in Russia, cars and trucks frequently bog down in the mud. Rather than constructing paved roads and highways, the Soviets, after consulting a noted theoretician, require that all vehicles carry a long, strong rope. According to the authorities, because there are always trees near the roads, one can obtain infinite force at the end of the rope, so any vehicle, no matter how heavy, may be pulled out of the mud. Is this true? How is it done? What should you do?

**HINT:** *Ropes may be pushed as well as pulled. The correct ropes envisioned by the Party are to be ideal in all respects.*

# Chapter 9

# Gravity and the Harmonic Oscillator

## The Balance of Power

A scale is balanced with a partially filled glass of water on one side. Will the balance be upset if you put your finger into the water carefully, without touching the glass?

1.

HINT:

This and related problems were resolved in ancient Syracuse.

# Vanya and the Water Bucket (A Matter of Location)

2.

On a cold winter morning, Vanya goes down to the river and brings a bucket of cold water into the entrance hall of his *izba* (log cabin). The traditional Russian winter water supply has been for many centuries a hole in the ice of a frozen river. Vanya is proud of the fact that he has not spilled any water, although he has filled the bucket right up to the brim. In fact, there are pieces of ice floating in the bucket, and the above-water portions of these pieces are above the brim. Vanya's mother tells him to remove the ice and throw it out, because the bucket will overflow when the ice melts. Is this precaution necessary?

HINT: *Archimedes' Law is not our only concern here. In fact, the correct answer is not the same for Russia as it would be for the U. S.*

# Towing the Barge

3.

This problem was first posed in the late nineteenth century to applicants for admission to the Central Paris Engineering Academy. A barge is moved up the Rhone River from Marseille to Lyon. Lyon is at a higher altitude—170 meters higher—than Marseille. You are required to calculate the total work needed to move the barge from Marseille to Lyon, and you are provided with data for the total work done against the resistance of the moving water during this trip. Everyone knows that when you calculate the work needed to drag any body up an inclined plane, you need to add the change $mgh$ in potential energy to the work done against friction. Should you similarly add in this case the work $mgh$ (where $m$ is the mass of the barge, $g$ the gravitational acceleration, and $h$ the increase in altitude) necessary to increase the potential energy of the barge? After all, the barge will be 170 meters higher up.

HINT:
Remember Archimedes!

## Beam Me Up, Scotty!

4.

To minimize "g-forces"—the forces due to high accelerations in spaceships during launching—it is proposed that astronauts be immersed in water tanks. This scheme is often used in science fiction. Floating in the water, the astronauts would be essentially weightless and not suffer the effects of high accelerations. Will this work?

**HINT:** *Consider whether the forces involved are of the same type.*

## Fixing the Clock (A Grave Matter)

5.

Christian Huygens, a seventeenth-century Dutch mathematician, astronomer, physicist, and creator of the wave theory of light, also invented and built the first accurate pendulum clocks. He sent one of these clocks from France, where it was designed and constructed, to a friend in North Africa. His friend, in writing a letter of thanks, mentioned that the clock was not as accurate as expected; in fact, it was running slow. Huygens immediately realized what the problem was and by return mail told his friend how to adjust the clock. What was Huygens's advice, and what was the problem?

**HINT:** *How do pendulum clocks work?*

## The Clock in the Elevator

6.

An elevator operator records her time on the job by punching in and out on a time clock. This is a *very* old elevator, and the time clock, which is right in the elevator, uses a grandfather clock with a pendulum. The acceleration and deceleration of the elevator are perfectly smooth and constant, and have the same absolute value (smaller than the gravitational acceleration $g$) whether the elevator is rising or descending. If the operator is paid by the hour, is she overpaid or underpaid?

**HINT:**

*The period of the pendulum will be altered by changes in the effective gravitational acceleration, so that the apparent working hour on the elevator may be longer or shorter than an hour at the office.*

## The Candle on the Carousel

7.

Light a candle and put it inside a tall glass chimney, so that wind will not disturb the flame at all. Take this candle lantern and go for a ride on a carousel. In which direction will the flame lean?

**HINT:**

*If your mood is buoyant and you are not overcome by inertia, you should be able to solve this problem.*

# Why Does the Moon Cause Tides?

8.

It is common knowledge that inside a spaceship flying in space with the rockets turned off, there is no gravity. For instance, we have all seen pictures from the space shuttles in orbit with objects floating freely about the cabin. In that case, the gravitational acceleration of the Earth is canceled by the acceleration due to the orbital motion about the Earth. Now the Earth can be considered as well, in this framework, as an unpowered spaceship. Therefore, all the objects on the Earth's surface must be weightless with respect to the gravitational forces from the sun, the moon, and all other celestial bodies. Why, then, does the moon cause tides in our oceans and lakes?

**HINT:**

*As spaceships go, the Earth is a rather large one!*

# Gravitational Attraction at the Breakfast Table

9.

Boris has learned that all bodies, according to Newton's Law of Gravitation, are attracted to each other, and that the attractive force increases as they get closer together as the inverse square of the distance between them. His grandfather, sitting at the breakfast table, challenges Boris to find an example where the gravitational attraction force *decreases* as two bodies are brought together. Can you find such an example?

**HINT:**

*The equation for Newton's Gravitational Law is formulated, strictly speaking, only for two concentrated masses. You should also consider what Boris's grandfather may be looking at on the breakfast table.*

# Lights Out at Malakhovka (The Harmonic Oscillator)

10.

You are celebrating enthusiastically at the village festival in Malakhovka. All the dachas are lit up. In the meeting hall, the authorities have hung bright lightbulbs from the high ceiling by long, springy cords. The loud music sets the lightbulbs bouncing up and down, the cords stretching and contracting, the height of each bulb oscillating with time. You notice that each bulb appears brightest at the highest and lowest points, that is, at the end points of the oscillation. Is this true, or have you had too much vodka? The continued motion of the lightbulbs irritates you so much that you go to your dacha and get your trusty AK-47 to shoot them out (perhaps it *is* the vodka). When should you fire your weapon in order to have the best chance of hitting a bulb? What should you aim at?

**HINT:** *Think about the title of this problem and what is meant by the term* brightness.

# The Platinum Planet

## 11.

Our spaceship is in a low circular orbit around a nearby planet in a newly discovered star system. The cook puts dinner into the oven, sets the timer for 45 minutes, and everyone relaxes after the long voyage. The timer goes off after 45 minutes, and the cook notices that the ship is at exactly the same point in its orbit as it was when he put dinner into the oven. He mentions this to the crew, and the technology officer becomes quite excited, saying, "This is only slightly more than half the orbital period for the low orbit space station back on Earth!" The captain immediately plans a landing party, stating, "It looks like it's made of pure platinum. Let's go!" How did the captain come to this conclusion?

**HINT:** *The average density of the Earth is 5.5 g/cm$^3$. The density of platinum is approximately 4 times greater.*

# All Tunnels Lead to Moscow

12.

The lifeblood of the Soviet system, the paperwork from the bureaus, had to appear in Moscow, particularly the Kremlin, as quickly as possible. The sheer mass of the paperwork and the long distances involved caused considerable difficulty. Electronic communications were distrusted because of secrecy concerns. It was proposed to solve all these communication problems once and for all by the construction of absolutely straight evacuated tunnels directly through the Earth among the major cities of the Soviet Union and the Kremlin. The papers were to be sent in metal containers suspended inside the tunnel by the magnetic field of superconducting coils, so there would be no friction whatsoever. The bureaucrats from Vladivostok (about 10,000 km away, on the Pacific Ocean) hoped to be the fastest in delivering their reports to Moscow because they were almost on the opposite side of the Earth and their tunnel was supposed to be the steepest, so the tubes would have the greatest acceleration. Apparatchiks from Leningrad (now Saint-Petersburg which is 700 km away) hoped that they would be the fastest because they were much closer. Who had the better opportunity to deliver their reports faster? Consider the Earth to be a sphere with constant density.

**HINT:**

*What will the velocities of the mailed containers be in Moscow if they are just dropped in at the other end? Also recall that in full analogy with electrostatic fields, a massive spherical shell generates no gravitational forces at any point inside it.*

## Chapter 10

# Mechanics and Relativity

## Crossing Swords

**1.**

You notice, during the swordplay in a performance of *Macbeth*, that it is almost impossible to observe in detail the motion of the points of intersection between the swords. Careful observation of a single duel shows you that such points move at times with extremely high velocity, too fast at times for human eyesight. Can you explain this?

**HINT:**

*In order to simplify the problem, reduce it to the situation of two bars intersecting at an angle $2\alpha$, each bar moving perpendicular to itself with a velocity $v$. What is the velocity of the intersection point? Choose a symmetric frame of reference. The laboratory frame of reference appears to be such a frame of reference because the speed of both bars would, in that frame, be the same.*

# Special Relativity in the Tailor Shop (Sasha Exceeds the Speed of Light)

2.

Sasha has solved the previous problem in school. Now, sitting in his father's tailor shop and examining a large pair of scissors, he decides that he is able to violate the Principle of Special Relativity. The previous problem describes, he feels, the kinematics of scissor blades. The blades rotate about a pivot, and therefore the velocity of any point on a blade will be perpendicular to the axis of the blade. Now, the answer to problem #1 gives the velocity of the intersection point as $u = v/\sin\alpha$. It is clear (to Sasha, at least) that as the scissor closes and the angle $\alpha$ decreases, the denominator in the expression will approach zero, and the velocity $u$ of the cutting edge (the intersection of the scissor blades) will increase and approach infinity. As Masha enters the shop, Sasha, greatly excited, points out to her that he has discovered that the speed of light can be exceeded and the principle of special relativity violated.

In fact, the previous problem involved pure linear lateral motion on the part of the swords, and in this problem we have rotation. To be more precise than Sasha, consider the figure on the next page.

The blades pivot at the point $O$, and the cutting is done by the intersection point $M$. The distance $s$ from the pivot to the intersection of the blades (the distance from point $O$ to point $M$ in the figure) is given by the relationship $h=s\sin\alpha$, where $h$ is the

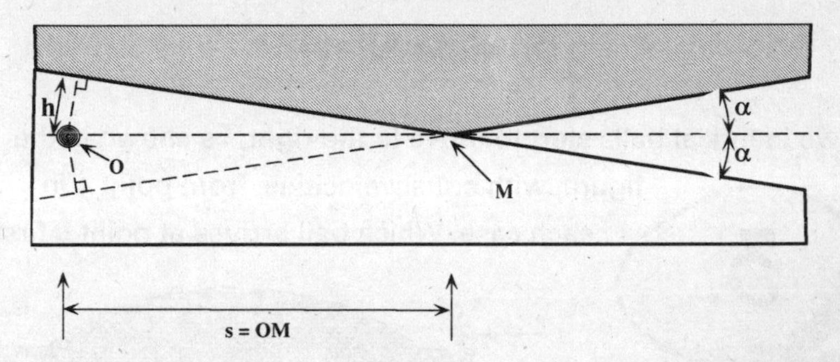


distance from the pivot to the blade edge and $\alpha$ is one-half the angle between the blade edges. Solving for $s$,

$$s = \frac{h}{\sin\alpha}$$

The derivative of this expression may be taken with regard to time. Then

$$v = \frac{ds}{dt} = \frac{d\left(\frac{h}{\sin\alpha}\right)}{dt} = -h\left[\frac{\cos\alpha}{\sin^2\alpha}\right]\frac{d\alpha}{dt} = h\omega\left[\frac{\cos\alpha}{\sin^2\alpha}\right]$$

For small angles, the answer can be further simplified, since $\cos\alpha \cong 1$ and $\sin\alpha \cong \alpha$ as $\alpha \to 0$:

$$v = h\omega\left[\frac{\cos\alpha}{\sin^2\alpha}\right] \cong \frac{h\omega}{\alpha^2}$$

The kinematics of scissor blades make the cutting edge move even faster, at small angles, than the intersection point of the "swords" in the previous problem! Has Sasha done it? Should Masha call the Nobel Prize Committee? (She doesn't think so.)

**HINT:**

*There are really two problems here. First, has the Principle of Special Relativity been violated? Even if you disagree, you should continue to examine this problem.*

## Bowling Alleys

3.

Two identical balls start to move to the right, as shown in the figure, with equal velocities, from point $A$ in each case. Which ball arrives at point $B$ first?

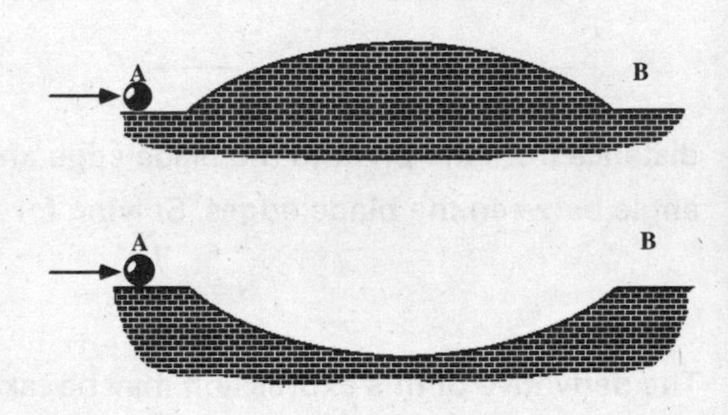


HINT:

*What can you state about the balls' velocities when they reach point B? Can you extend this idea?*

## Boris and the Locomotive

4.

Boris has just learned about instantaneous centers of rotation: any point that is at rest even instantaneously can be used to calculate moments and rotational velocities. His grandfather, who is watching a locomotive from a hillside where he and Boris are relaxing, remarks that the concept seems artificial and even misleading. He points out that the cylinder of the locomotive pushes a piston out, which in turn pushes a push rod backward, as shown in the figure.

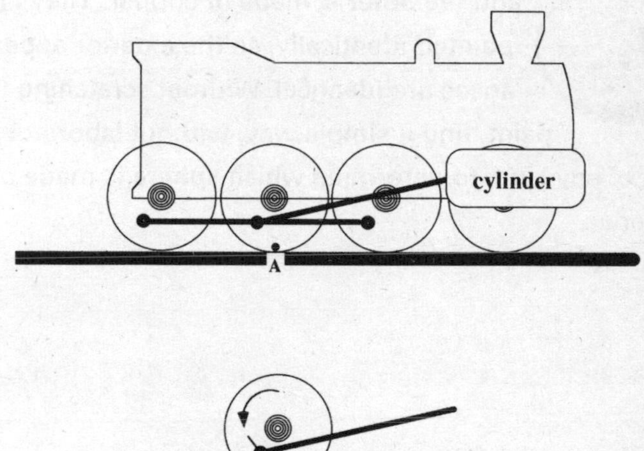


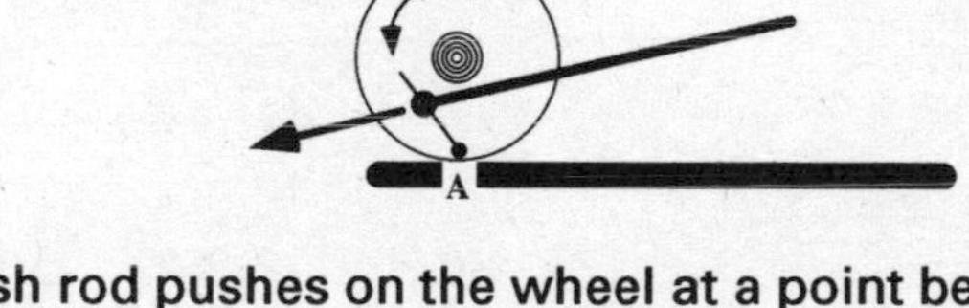


The push rod pushes on the wheel at a point between the axle and the point $A$ at which the wheel rides on the rail. Point $A$ can be used as an instantaneous center of rotation, and then the connecting rod creates a counterclockwise moment when it pushes on the wheel. Consequently, grandfather points out, the wheel must be rotating counterclockwise and the locomotive must move backward. Why, then, when the rod pushes on the wheel, does the locomotive move forward?

HINT:

*The analysis is correct as far as it goes. But the locomotive does indeed move forward.*

# The Ball Game

5.

Before you are two hollow spheres with exactly equal mass and outer diameter. One is made of aluminum and the other is made of copper. They are painted identically, so the exterior appearances are identical. Without scratching the paint, find a simple way, without laboratory equipment of any kind, to determine which sphere is made of which material.

**HINT:** *Ballplayers toss the ball around. Cheerleaders twirl batons. Who learns more mechanics?*

# A Game of Billiards (Visiting the Ergodic Theorem)

6.

In a billiard table with sides $a$ and $b = 2a$, a ball is launched from the middle of the long side. Find a launching angle (say, the angle $\phi$ with the long side) that guarantees that the ball will return to the point from which it was launched.

Assume the collisions with the sides are elastic (no energy loss) and remember that, unlike pool tables, billiard tables have no pockets.

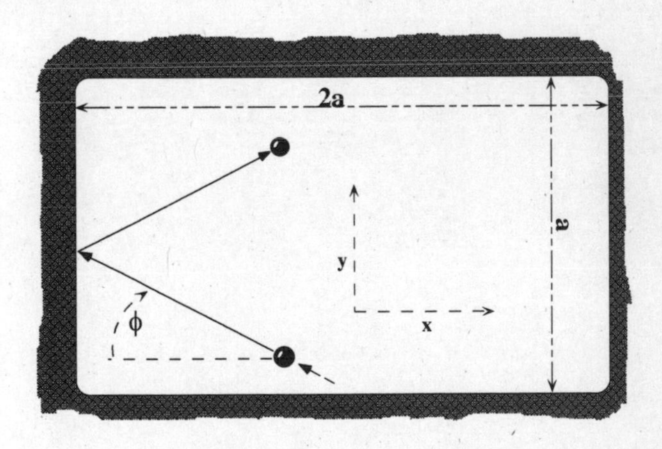


HINT:

*For elastic collisions, the velocity component of the ball parallel to the side remains unchanged.*

## Winter Fun

7.

Vanya is out in the snow with his sled. He climbs to the top of the hill, sits on his sled, and slides down. He tries higher hills in order to get a longer ride, but this appears to be a losing proposition. He climbs a new hill, 4 times as high as the previous one, hoping to get a ride that will last 4 times as long. The new hill is not straight but is geometrically similar to the previous one. How much longer will it take to sled down? The air resistance is negligible, and the friction between the snow and the sled does not depend on velocity.

**HINT:**

*A physical relation or equation must be independent of the system of units that is used. This statement is sometimes referred to as the Zeroth Law of Nature. Examine the physical variables in Vanya's case in the light of this law.*

## Bow and Arrow

**8.**

For any given bow, there is always an optimum arrow length—that is, an arrow that can be shot and that will fly the maximum distance. Explain why this is so.

**HINT:** *There is a well-known theorem that applies to continuous functions: if a non-negative function is continuous over a given interval and is equal to zero at the ends of the interval, it must have at least one maximum within the interval. As we mentioned in the solution to problem #9, chapter 9, Lev Landau, one of the great theoretical physicists, applied this principle very broadly. You can apply this principle specifically to the arrow. Remember, it must be possible to aim the arrow.*

## High Technology in Russia

**9.**

Ivanov is moving his boss's safe, which is extremely large and heavy (it is used to store Party secrets), from one office to another. (Due to the general paranoia in the Soviet Union, such a safe was found in almost every office. One of us [Y.C.] used his safe to keep reagent grade alcohol—necessary in the laboratory—secure because otherwise the bottle would be stolen and immediately consumed.) Some of the "secrets" are fragile, so the safe must be moved carefully; the safe cannot jerk up and down as it is moved. The traditional method for achieving this is to put the

safe on a heavy metal plate and roll the plate on logs, choosing the roundest logs possible.

However, round logs are very scarce at this time, and Ivanov is in trouble. The boss's secretary (who is also a mathematician) tells Ivanov that it is not necessary for the logs to be round, that a perfectly level ride can be achieved, with the plate perfectly parallel to the floor, with noncircular cross sections. Is the boss's secretary teasing Ivanov, or is this really possible? (Ethnic jokes were not unknown in such places!)

HINT: *The distance between the plate and the floor must remain constant. However nothing was specified for the horizontal position or velocity, and these may change quite suddenly if you wish!*

# The Jack-in-the-Box (An Exercise in Renormalization)

## 10.

If you depress the head of the Jack and release it suddenly, it will pop up because of the spring between the Jack and the bottom of the box. How much force is required to depress the Jack so that both the Jack and the box will jump up into the air? The masses of the Jack and the box are known, and the mass of the spring (a typical linear spring) is negligible.

**HINT:**

*Consider the Jack-in-the-box as two masses, m (Jack) and M (the box, resting on a table), connected vertically by a spring, the box on the bottom, the Jack on the top. The minimum force necessary will cause the box to rise to a negligible (infinitesimal or zero) height above the table. Because the force is the minimum required, the velocities of both the Jack and the box will be zero at that instant. This is a very special moment, because the sum of the forces on the box and the total energy of the system take on particularly simple forms at that time.*

# Chapter 11

# Geometry and Numbers

## The Chicken from Minsk versus the Information Superhighway (or Why Did the Chicken Cross the...?) —A Fable of the New Russia

1.

A fiber-optic cable completely encircles the Earth, by chance running through a new, privately owned chicken farm on the outskirts of Minsk. The chickens refuse to walk or fly over the cable and will only pass under it. Clearly, the cable must be raised off the ground by 1 foot if the chickens are to survive. For technical and bureaucratic reasons, the cable must then be raised 1 foot higher everywhere, around the entire circumference of the Earth (the cable is thick and fragile and cannot be bent or twisted, even slightly). The farmer, exercising his newly won individual rights (no more USSR!), refuses to permit the cable to cross his land unless it is raised. The bureaucrat in charge of the project is a holdover from the old days. He maliciously agrees to raise the cable only if the farmer agrees to pay for *all* of the additional cable, at the rate of $1 dollar (U.S.) per foot. The farmer agrees, provided that the government will pay for the additional supporting structures. How much will the chicken farmer pay?

**HINT:** *Use circular reasoning.*

## Family Fun with Mathematicians

2.

One day two mathematicians, Igor and Pavel, meet in the street. "How are you? How are your sons?" asks Igor. "You have three sons as I remember, don't you? But I have forgotten their ages." "Yes, I do have three sons," replies Pavel. "The product of their ages is equal to 36." Looking around and then pointing to a nearby house, Pavel says, "The sum of their ages is equal to the number of windows in the building over there." Igor thinks for a minute and then responds, "Listen, Pavel, I cannot find the ages of your sons." "Oh, I am very sorry," says Pavel; "I forgot to tell you that my oldest son has red hair." Now Igor is able to find the ages of the brothers. Can you do it?

**HINT:** *Pay careful attention to what is said and who says it!*

## Some Numbers!

Prove that the number $(n^2\text{-}1)$, where $n$ is a prime number other than 2 or 3, is divisible by 24.

**HINT:** $(n^2\text{-}1) = (n\text{-}1)(n+1)$. *Do not forget the other conditions of the problem.*

## The Root of All Our Problems

What is the limit of the expression

$$x_n = \sqrt{2+\sqrt{2+\sqrt{2+\sqrt{2+\ldots\sqrt{2+\sqrt{2}}}}}}$$

when the number of square roots $n$ tend to infinity?

**HINT:** *Dig deeply into the roots of your problem.*

## Scaling the Pythagorean Theorem

Prove the Pythagorean Theorem by using scaling.

**HINT:** *Areas of similar triangles are proportional to the squares of similar sides.*

# Stretching Olya's Mind (or A Stretch of Olya's Imagination)

6.

Olya has returned from the movies (see chapter 2, problem #3) and is doing her homework, which requires her to calculate the area enclosed by an ellipse. She cannot find her textbook (she thinks she left it at the movies) and cannot remember the formula. She does remember the formula for the area of a circle, which is $S = \pi r^2$. In desperation, she draws the circle in her drafting notebook, on pages that have fine grids drawn on them dividing the surface into many small squares. Petia (who is actually hiding her textbook) teases her by saying, "You must stretch your mind." At that point, Olya understands how to solve the problem. How does she do it?

HINT:
*An ellipse can be obtained from a circle by uniform stretching in one direction.*

# The Problem That Didn't Fool Von Neumann

7.

Go back to problem #10 in chapter 2, "The Crazy Dog." This is a version of what is generally referred to as "The Problem That Fooled John Von Neumann." We pointed out there that the problem was a trick question that could be solved very quickly by looking at the time the dog had to run. When the problem was proposed to Von Neumann by his friend and colleague Stanislaw Ulam, he solved it immediately, and Ulam replied, "Oh, you saw the trick." Von Neumann's reply was, "Trick? What trick? I just summed the series." The general assumption is made that the series solution is very complicated. In fact, it is not. Can you do it?

HINT:

*After the dog completes the first cycle—that is, he runs from Misha to Tisha and returns—the whole situation is exactly as it was at the beginning, but rescaled by a certain factor, as the bicyclists are now closer together. To solve the problem, you will need only to find that factor.*

# The Party Secretary's Fence

8.

A fence is to be constructed around the Party Secretary's dacha (vacation house), and the top of the fence must have the shape described in the figure.

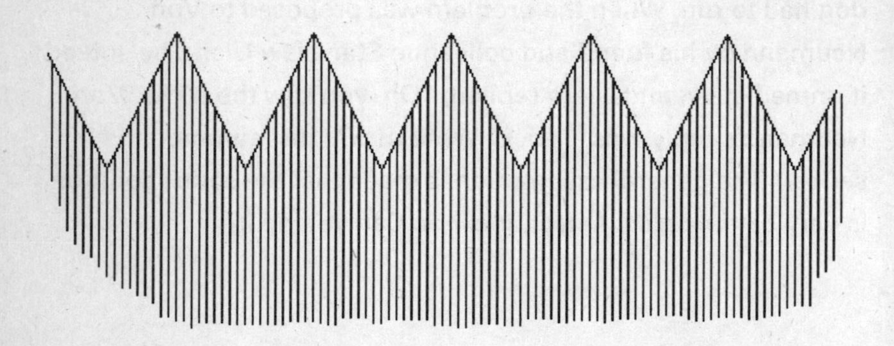

In charge of fence erection is the mathematician Igor (his second job is construction). He intended to give the drawings to Pavel, the carpenter, but while he looked for Pavel, the Party Secretary's goat came along and ate the drawings. Fortunately, Pavel is also a moonlighting mathematician, so Igor wrote down an equation describing the top of the fence. Can you do it? (Because the Party Secretary's dacha is so large, the length of the fence should be considered to be infinite.)

**HINT:** *There is more than one answer.*

## What Color Was That Bear? (A Lesson in Non-Euclidean Geometry)

**9.**

A camper leaves his tent and walks South for 2 kilometers. Then he turns to the West and walks 5 kilometers. Finally he turns to the North and walks 2 kilometers. How far from his tent is he?

**HINT:**

*There are situations when an equality is not the correct answer. One way or the other, you will need warm clothing for this one!*

## Identity Crisis

**10.**

It is well known that the Earth is a magnetic dipole with a North Pole and a South Pole. It is also known that these poles are situated not far from the geographical poles. Which of the magnetic poles is near the North geographic pole?

**HINT:**

*The concepts of the North and northerly directions were accepted in Europe well before the magnetic compass was brought there from China.*

# Solutions to Chapter 1

## WARMING UP

**1. Natural Childbirth:** Triplets! They are two members of a triple birth. (Or quadruplets, quintuplets . . .) We told you to pay attention! You would have had an easier time with this one if we had asked you to define what twins are.

**2. The Dumb Parrot (or The Problem with Mathematicians):** There are a number of solutions to this problem. Here are a few:

(a) The bird is deaf (in physical terms, there is no initial condition). This is the most obvious solution. The bird will repeat every word he hears, but he cannot hear anything.

(b) Since the proprietor did not say *when* the bird will repeat what he hears, another answer is that the bird will repeat every word—in a few years. (The mathematician's solution has been translated along the $t$ axis.)

(c) The highly intelligent bird may well have ignored his new owner, who was an extremely boring conversationalist. After all, would an intelligent person speak to the parrot just to have it repeat the words?

(d) The customer may be lying. (His wife found out how much the bird cost and forced him to return it.)

There are many solutions even more ridiculous than these. The situation arises from application of rigorous mathematical language to everyday usage, which

is imprecise. The use of mathematical rigor in common communication would be maddeningly inconvenient. In everyday language, much is tacitly implied. The scientist lives between these extremes, and needs to check all definitions and descriptions to be certain of his or her understanding.

**3. A Question of Art:** If you draw a square matrix or table with names as columns and hair color as rows (or vice versa), diagonal elements are prohibited, since then the hair color and name would be the same. We know that White responded to the person with black hair, so he cannot have black hair. He must have red hair, and this leaves black and white for Black and Red. Black must have white hair and Red, the artist, must have black hair.

**4. To Eat or Sleep, That Is the Question!:** Because you cannot simultaneously sleep and eat, the time from your last sleep and your last meal must differ. You must do the last thing you did two weeks before; if you slept and then ate before your ordeal, you must eat first afterward. (Considering the length of the food lines in the former Soviet Union, this could be a very real problem. Would you sleep first or eat first before going to the store?)

**5. The Knights and the Pages:** The following sample solution will work only if all the pages get along and the knights refrain from killing each other during the crossing. This solution is not unique. There are others. (See facing page.)

**6. More Knights and Pages:** There is no solution unless one of the four pages is sacrificed. (In medieval times, this was not a problem.) There is no way of getting four or more pairs of knights and pages across without at least one of the pages having difficulty breathing at some point in the process.

There are only three possible first moves: the boat carries two knights, a page and a knight, or two pages. The first possibility creates immediate breathing problems for the two pages who are left behind by their masters. The second would allow only the knight to return; this is followed by a pair of pages crossing, one page returning, and then the remaining two pages crossing over. At that point all the pages have crossed and all four knights remain, a situation similar to step 4 in the solution to the previous problem. This will also be the inevitable result if in the first move, the two pages cross, the only remaining possibility: a page returns, two pages cross, and the cycle is repeated one more time. In fact, after the second crossing, both the second and third possibilities lead to the identical position. Consequently, the position with four pages across and four knights remaining is inevitable.

However, this position shows us that no solution is possible. One of the pages has to return the boat. The knights cannot cross now. If two knights cross, the third page on the other side is in trouble, and if the knight-page pair crosses, all three of the waiting pages on the other side are in jeopardy.

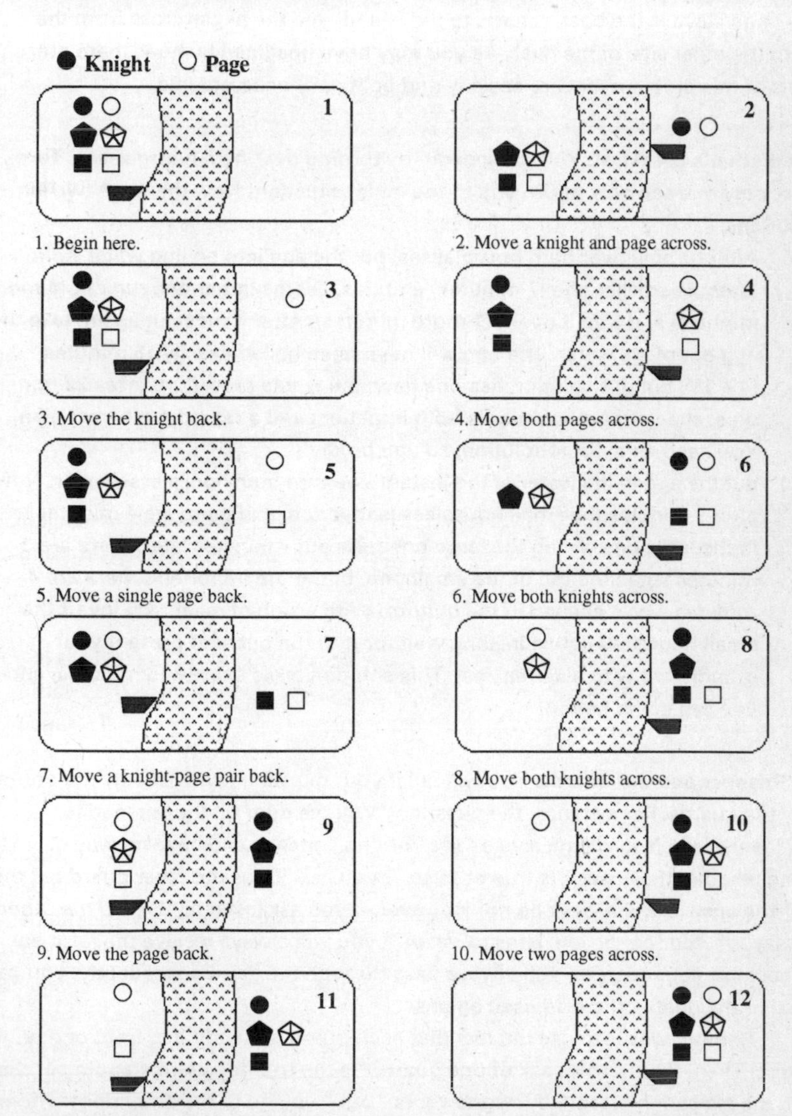


1. Begin here.
2. Move a knight and page across.
3. Move the knight back.
4. Move both pages across.
5. Move a single page back.
6. Move both knights across.
7. Move a knight-page pair back.
8. Move both knights across.
9. Move the page back.
10. Move two pages across.
11. Move one page back.
12. Move two pages across. All done!

**7. Yet More Knights and Pages: No Man Is an Island:** The island will solve the problem with any number of knights and pages. First, move all the pages to the island. Now each page escorts his master to the other side of the river and returns to the island. The process continues until all the knights have crossed. After the final crossing the page stays in the boat, returns to the island, and the pages cross from the island to the other side of the river. As you may have imagined by now, there are versions of this problem that are socially and politically unacceptable.

**8. Grandfather's Breakfast:** You *must* begin by turning over *both* hourglasses. Turning over only one can only return you to the initial situation! Now there are two simple solutions:

(a) After turning over both hourglasses, put the egg into boiling water *right after* the smaller one (7 minutes) empties. Let the larger one run out (4 more minutes) and turn it over (11 more minutes). After it runs out again, take the egg out of the water. The egg will have been boiled exactly 15 minutes (4 + 11), but this solution has one flaw: the whole procedure takes 22 minutes, and your grandfather is both impatient and a retired mathematician. You need an optimal solution, so you boldly

(b) put the egg in the water at the instant you turn both hourglasses over. Now, after 7 minutes the small hourglass is inverted, and there are 4 minutes left in the large one. When the large one runs out 4 minutes later, there are 3 minutes left in the top of the small one, but, more important, *there are 4 minutes worth of sand in the bottom!* With a sigh of relief, you invert the small hourglass at this instant, wait for it to run out, and present your grandfather with his breakfast. This solution takes exactly 15 minutes, and you can do no better.

**9. The Prisoner and the Guards:** You will fail if your question involves only the doors or only the guards. For instance, the question "Will the door on the left lead to death?" is useless. You will receive a "yes" or "no" answer and have no way of knowing whether the answer is true or false. If you ask "Does the other guard tell the truth?" the answer will *always* be no! However, if you ask instead, "*Would the other guard say the door on the left leads to death?*" you will always receive the false answer, because your question will always pass through the liar. Consequently, you can proceed to the door you are advised against.

Another way is to use the fact that each guard is standing in front of one of the doors. Then, for instance, ask of one guard, "Is the truthful guard standing at the door to the execution block?" If the answer is "yes," you go to the *other* door. There are really only two circumstances: the truthful guard is either standing at the door to life or at the door to death.

**10. Ivanov and the Clock:** Just before high noon. There is only one time period over which this clock will strike once even *three* times in a row at half-hour intervals, and that is at 12:30, 1:00, and 1:30. The trials of Ivanov *must* include this time period. How to account for the fourth time? Ivanov must have opened the door just as the clock was finishing its 12 strikes for noon, just before the final strike. Because of the furnishings and soundproof door he would not have heard the first 11. Could the fourth time be 2:00 P.M.? No, because this time Ivanov had left the door open and would hear both strikes.

# Solutions to Chapter 2

## MORE WARM-UPS

**1. Shopping with Boris and Marina:** 25 kopeks. If Boris had had 2 or more kopeks, then he and Marina would have had sufficient funds, since Marina was only 2 kopeks short. So Boris had 1 or 0 kopeks. If 1, then because he was 24 kopeks short, the price is 25, and Marina is indeed 2 kopeks short. You may object to this answer, since if Boris had 0 kopeks, the price would be 24 kopeks and Marina would have 22 kopeks. However, we will assume that Boris was not conning Marina and that he did have something in his pocket.

**2. Shopping with Boris and Marina, Part II:** Yes, your suspicions are quite correct. Boris has no money, none at all. His sister has enough money for two chocolate bars. You may now want, in the light of this revelation of Boris's character, to reconsider your answer to the previous problem.

**3. At the Movies:** They will meet 1,260 days later. The number of days between meetings must be divisible by 4, 5, 7, and 9. However, none of these numbers has a common factor, so the minimum number satisfying the above condition is the product, $4 \times 5 \times 7 \times 9 = 1{,}260$. Since high school in Russia lasts three years, this schedule was well planned. None of them were caught.

**4. A Drinking Problem:** Look at the figure.
The amounts of vodka in each container are:

a. (12, 0, 0) initial condition: 12-liter bucket filled, others empty.
b. (4, 8, 0) fill the 8-liter bottle fully from the 12-liter bucket.
c. (4, 3, 5) fill the 5-liter bottle from the 8-liter bottle.
d. (9, 3, 0) empty the 5-liter bottle into the 12-liter bucket.

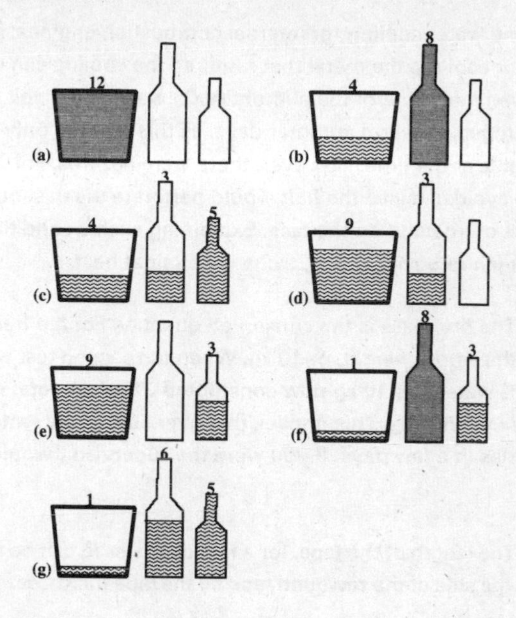


e. (9, 0, 3) pour the 3 liters of vodka from the 8-liter bottle into the 5-liter bottle.
f. (1, 8, 3) fill up the 8-liter bottle from the 12-liter bucket, leaving 1 liter behind.
g. (1, 6, 5) fill the 5-liter bottle by decanting 2 liters from the 8-liter bottle, leaving 6 liters behind.

You now have 6 liters in the 8-liter bottle. Empty the 5-liter bottle into the 12-liter bucket and you have (6, 6, 0) 6 liters for Vassily and 6 for Pyotr.

**5. The Caterpillar:** It will take 6.5 days. Beginning at sunrise on the first day, when the caterpillar is resting on the ground, it will reach, by sunset of the first day, 12 feet; by each successive sunset, 18, 24, 30, 36, 42, and 48 feet.

**6. Washing Your Face:** The boiling point of liquid nitrogen (at 1 atmosphere pressure) is some 230° C below your skin temperature. The "extreme" heat of the skin quickly boils the liquid in contact with it and a thin gaseous layer forms between the skin and the cold liquid. The gaseous layer serves to insulate the skin from the cold liquid. If you try the water droplet in the frying pan, you will note what seems to be contradictory behavior: if the pan is hotter, the water droplet lasts longer. There is, of course, no contradiction; the high heat makes a better gaseous layer and slows down the conduction of heat to the droplet. The gas layer presents serious problems in the design of boilers and water cooling for internal combustion engines; if it forms where the water is used for cooling, the metal that required the cooling can overheat and burn through or even melt. One of the authors (Y.C.) was in the habit of washing his face with liquid nitrogen on warm summer days, as this was the only form of air conditioning available at the time. However, there were hazards. For instance, the eyebrows must be avoided, since the hair would perforate the gaseous layer, resulting in a severe case of frostbite on the face. Explaining such a condition to your superior was, at least in the Soviet Union, a chancy affair at best.

**7. Buying Berries:** The dry mass is the conserved quantity. For the fresh berries it amounts to 1% of the whole weight, or 10 kg. When the second test was made and the berries had 98% water, the 10 kg now constituted 2% of the total mass. Therefore, the total mass was then 500 kg. This implies that over 50% of the water had evaporated from the berries in a few days. If you were the supervisor, would you believe this story?

**8. The Videotape:** The length of the tape, for a reel diameter $R$, can be found by dividing the surface area of the side of the rewound tape by the tape thickness:

$$L = \pi(R^2 - r^2)/\delta,$$

where $r$ is the bobbin diameter and $\delta$ is the tape thickness. Because $R$ is much larger than $r$ this relation becomes

$$L \approx \pi R^2/\delta.$$

If the bobbin rotates at constant angular velocity (e.g., 120 rpm), the outer radius will grow at a constant rate and therefore the length of tape rewound will grow as the square of the time. The first quarter of the total length will then take one-half of the total time. Note that the answer for the *last* quarter is quite different! Also note that the situation is different when the videotape is playing: now the *tape* velocity is constant and the angular velocities of both reel (accelerating) and bobbin (decelerating) will vary with time.

**9. Adventures on the Moscow Subway:** Suppose the clockwise trains arrive exactly on the hour and the counterclockwise trains arrive 12 minutes later. If Boris arrives between any hour and 12 minutes past the hour, the next arrival will be a counterclockwise train. However, if he arrives after 12 past the hour, the next arrival will be a clockwise train. For 20% of the time (12 minutes of every hour), Boris will board a counterclockwise train; for the other 80%, a clockwise train. If Boris arrives at random times, he will end up with clockwise trains 80% of the time. That is to say, there are 4:1 odds in this case that Boris will catch the clockwise train! As long as the trains

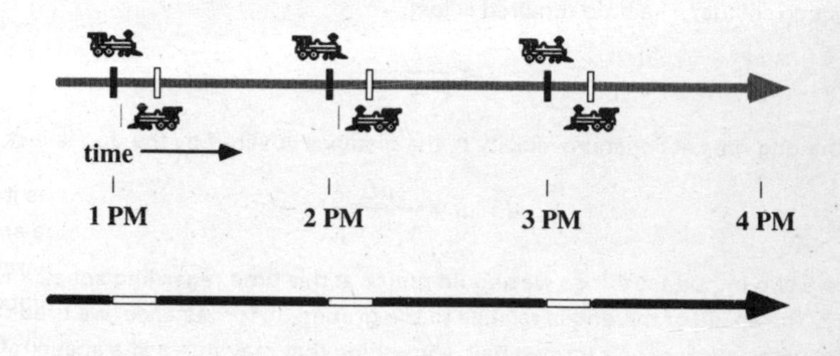


have a regular schedule with regular intervals between arrivals, such an effect will occur. The time axis will be divided up into time spans and the length of the "counterclockwise" and "clockwise" spans will depend on the phase shift, that is, the time interval between the sets of point. As long as the phase shift is not zero or half of the scheduled time intervals for the trains, the time spans will be unequal in length. Therefore, there will generally be a difference in the probability of catching the two types of train.

Does this result seem counter to your intuition? If so, your intuition was wrong. Random and regular events must not be confused. For Boris, the arrival of the train is a random event because he arrives at random moments. Both the clockwise and counterclockwise arrivals are random events, from Boris's point of view. At that point he concluded that these events are *uncorrelated* random events, which is most certainly wrong. His arrival is a random event superimposed on the highly regular arrivals of the trains. Recognition of these facts makes the situation clear.

We have used 1-hour intervals because we have become accustomed to the Boston transit system. In Moscow, the typical interval during rush hour was less than 1 minute. The organization of the transit system was along military lines, with military discipline.

**10. The Crazy Dog (or The Problem That Did Not Fool John Von Neumann):** The repetition of the dog's route is, of course, the problem, as the bicyclists move closer and closer. The result of considering the details of the dog's route is an infinite series.

However, look at the initial and final situations. How long does it take for the bicyclists to collide? The time required is just

$$t = \frac{L}{v_1 + v_2}$$

Since the dog runs at constant velocity $u$, the distance covered by the dog is just

$$d = ut = \frac{uL}{v_1 + v_2}$$

There are two important things we should notice at this time regarding speed and velocity. The speed of the dog is relative to the ground. If, for instance, we tried this problem with an insect or a thrown ball, something that may move at a speed relative to the bicyclists, we'd have to take the bicyclists' speeds into consideration when we calculated the total distance traveled by the third object. We would have to add the velocity of the object to the bicyclist's velocity to obtain velocity relative to the ground. You may also notice that we stated speeds, not velocities. The dog's *speed* is constant, but its velocity changes each time it meets one of the cyclists as it changes direction. Velocity is a vector quantity that specifies both speed and a direction. Always be aware of the difference between speed and velocity!

This problem has interesting scientific and historical roots. The dog, frantically racing back and forth, may as well be a photon or some other particle connected with a field interaction, being exchanged between two other particles approaching each other. The historical roots go back to a discussion some 50 years ago between two famous mathematicians, Stanislaw Ulam and John Von Neumann. Involving a fly rather than a dog, it is referred to as "the problem that fooled John Von Neumann" (see *Mathematics Teacher,* November 1991). In our opinion it did not fool Von Neumann at all, as we shall prove to you in a later chapter ("Geometry and Numbers").

# Solutions to Chapter 3

## NIKOLAI'S MISERABLE BUSINESS TRIP

**1. Cooling the Tea:** For *any* mechanism of heat transfer, the heat exchange between two bodies will be greater if the temperature difference is greater. In this case, the "second body" is the air around the tea, and the heat loss from the tea will be greater when the tea is hotter. Putting the sugar in the tea immediately will drop the temperature, due to the endothermic reaction (see hint), and therefore cut down the heat loss. Thus the faster method is to delay the addition of sugar.

**2. Luxurious Accommodations:** Cut the third ring. You now have three pieces of the chain, 1, 2, and 4 links long. For the first night, pay 1 link. For the second, give the innkeeper the string of 2 links and take back 1. For the third, give the innkeeper the single link. For the fourth, give the innkeeper the 4-link chain and take back 1 and 2 links. The remaining three days are handled in the same way as the first three. (The ability to solve such problems was once necessary to survival in Russia.)

**3. The Worst Case in Vishny-Volochok:** 45 trials. There is only one trick here: if you have $n$ keys and you know that one of them will open the door, you need try at worst $n-1$ of these keys, since if the first $n-1$ keys fail, the remaining key is the correct key. No more than 9 trials are necessary for the first key, since if the first 9 fail, you know the remaining key is the correct key to open the door. Thus the worst case is $9 + 8 + 7 + 6 + 5 + 4 + 3 + 2 + 1 + 0 = 45$.

**4. Another Miserable Night in Vishny-Volochok:** Using both heaters in each cup is faster. Remember that the energy not only is spent on heating the water but also must be wasted on heat losses. The same amount of heat will be needed to heat the water to boiling from room temperature. Only the heat losses while the water heats up will be different. Whatever the general mathematical form for the heat losses, the rate of heat loss (that is, heat loss per unit time) should depend almost totally on the temperature difference between the cup and the room. Therefore, if the cup spends a greater time in any temperature interval, the heat loss should be greater. The more quickly the cup is heated, the less heat will be lost. If the cup is heated by two heaters, the energy loss will be smaller than if it is heated by one. For sequential heating by two heaters, the total energy required will be less and therefore less time will be required.

**5. Nikolai's Welcome Home:** Nikolai is innocent, at least in this case. In *both* cases, Vaska's claws will make straight lines on Nikolai's luggage. In both cases we are interested only in the horizontal component of Vaska's velocity. Before he lands on the luggage, the horizontal component will be a constant if aerodynamic drag is negligible (at cat velocities, this is true) and perpendicular to the constant luggage velocity. The only acceleration acting on the cat after he has leaped is gravitational, and this is *not* in the horizontal plane. Now use the luggage as your frame of reference. This is a Galilean frame and the cat's velocity must also be uniform and constant relative to the luggage. Then the cat lands on the flat, horizontal surface of Nikolai's luggage. If friction is negligible, the cat continues across the luggage at some constant angle, making straight gouges. If friction cannot be neglected, your first reaction may be that if the cat is decelerated by friction the angle must change, because his velocity changes after he lands on the luggage and the gouges will then be curved, as Nina thought. However, if you think that, you have deserted your reference frame. The friction force has the direction *exactly opposite* to that of the cat's velocity *relative to the luggage*, and therefore *only* the magnitude, but not the direction, of the relative velocity will change. Therefore, for both cases the gouges are *straight lines* at a constant angle to the direction of motion of the conveyor. It is important that aerodynamic friction be negligible (remember the hint?) because this friction depends on Vaska's speed relative to the air, not the luggage, and this friction force can change the shape of the gouges. (Cats, particularly Russian cats, care very little about the laws of aerodynamics.) Once Nikolai has explained this to Nina, he is forgiven.

If you would like a more quantitative answer, the angle $\alpha$ of the cat's velocity $\mathbf{u}$ to that of the conveyor belt's motion is given by

$$\tan \alpha = \frac{u}{v}$$

and the gouges will be straight and have the angle $\alpha$ to the direction of motion of the luggage on the conveyor.

This problem is intended to make you think about frames of reference in general. Turn to the next chapter to learn more.

# Solutions to Chapter 4

## FRAMES OF REFERENCE

**1. A Lesson in Aerial Combat:** An artillery shell has a very large initial velocity with respect to its "launcher." A missile begins with no initial velocity at all with respect to the airplane. Just before (or simultaneous with) the firing of the rocket motor, it is detached from the airplane. At this instant the missile's initial velocity with respect to the Earth (and the atmosphere around it) is the same as that of the airplane. However, the airplane is moving forward—that is, it is pointed in the proper direction; the missile, however, being pointed backward, is actually traveling backward in the air and feels the flow of air in a direction opposite to the usual direction of motion. Now the fins on the missile are designed to keep it moving forward only. The fins then turn the missile forward. Anyone who has played badminton has seen this happen: the skirt of the shuttlecock flips it over after it is hit by the racquet. In short, because of the fins, any backward motion of the missile is unstable and it will always turn and face forward in the airstream. This happens quickly at combat speeds, and when the rocket motor ignites, the missile accelerates and hits the plane that launched it. (We should note that the designers of this weapons system were never heard from again; at that time in Russia, the usual penalty for failed weapons systems was the firing squad or life in the gulag.)

**2. Old Man Mazay Rows for His Vodka:** Old Man Mazay will need exactly 30 minutes to get his vodka bottle back, if nobody swims out from the riverbank and gets to it first—a likely event in Russia. This answer does not depend on the rowing speed or current, only that they do not change. If we take the flowing water as our frame of reference, it is clear that his return time should be identical to the time elapsed since he dropped the bottle. Suppose the bottle had been dropped in still water, near a buoy. Clearly, all Old Man Mazay would have to do, in that reference frame, would be to retrace his course to the buoy, taking exactly a half-hour. The same should apply in moving water. Although the bottle is standing still in the river reference frame, it is moving at 2 miles per hour relative to the Earth. Therefore, it will be 2 miles away (downriver) from the bridge when it is picked up a half-hour later, which is 1 hour from the moment it was dropped into the water. You can see the advantage of this approach by solving the same problem in the reference frame that takes the Earth as the fixed point; now the redundant data we provided does make the problem more confusing.

**3. Fasten Your Seat Belt!**: Use a Galilean transformation to explain this situation. My plane was moving at a velocity $\mathbf{v}_0$ relative to the air and to the Earth (remember, there was no wind!), and the other plane moved at velocity $\mathbf{v}$. Since I was the observer, the reference frame should be fixed to my airplane. In this reference frame, the smaller plane should have a new reference velocity $\mathbf{v}'$, and the three velocities can be related by the Galilean transformation $\mathbf{v} = \mathbf{v}_0 + \mathbf{v}'$. The observed velocity $\mathbf{v}'$ of the small plane is then given by

$$\mathbf{v}' = \mathbf{v} - \mathbf{v}_0$$

Since $\mathbf{v}$ was parallel to the fuselage of the small airplane, the observed velocity now has a different direction, which created the puzzling effect I observed. The small airplane is moving through the air as it should, parallel to the fuselage and not slipping sideways. However, the velocity in *my* reference frame will not be parallel to the small airplane's fuselage unless both aircraft are moving in exactly the same direction relative to the Earth. If there is even a small angle, the velocity $\mathbf{v}'$ will have a sideways component and the plane will appear to be slipping sideways.

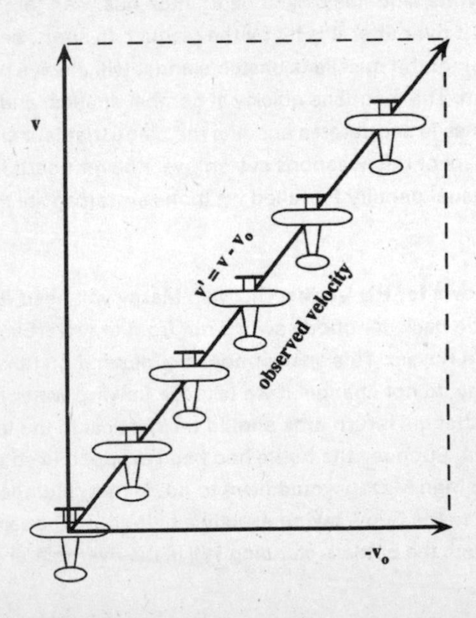

**4. Explosions:** The shape is determined by the relative motion of the balls, which does not depend on the reference frame. Immediately after the bag explodes, the geometric shape described by their positions is that of the surface of a sphere with uniformly growing radius $v_0 t$, where $v_0$ is the initial speed of the balls. In the presence of a gravitational field the growing sphere will remain, but the center of the sphere will fall with acceleration $g$. You can arrive at this conclusion most directly by using a non-inertial frame of reference—in particular, the center of mass of the balls, which must fall with the constant acceleration $g$. Or you can add this acceleration to the motion of each and every ball, and come to the same conclusion. The term with the gravitational acceleration will disappear whenever we are concerned only with the relative motion of the balls. The growing sphere should remain intact under the influence of any *uniform* force field. Aerial fireworks—for example, "star shells"—exhibit this effect because in the aerial explosion, the sparks initially fly outward with high speed, dying out before the effects of air resistance become significant.

**5. Rainy Day on the Carousel:** We want the dry cylinder to be as wide as possible, and that means that the handle of the umbrella must be parallel to the raindrop velocity vector relative *to you*, not the Earth. You are moving with velocity $\omega R$ relative to the Earth (and the center of the carousel), and the velocity of the raindrops relative to you will be the velocity of the raindrops relative to the Earth with your velocity subtracted by the usual vector method, as shown in the figure. Therefore, the umbrella must be tilted forward, with the angle of tilt $\alpha$ given by $\tan\alpha = \omega R / v$.

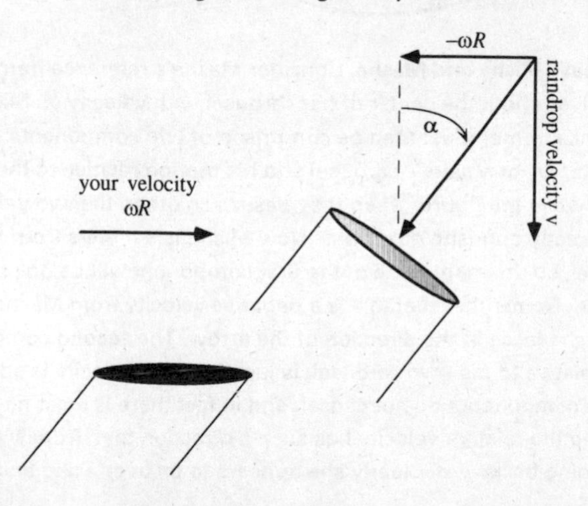


**6. Misha and Tisha at the Carnival:** They are on the carousel (riding, for example, side by side) or possibly a Ferris wheel. Look at the figure! They have the same angular velocity. Because Tisha is riding on the outside, his radius from the center of rotation is greater, and therefore he will have a greater linear velocity. The peculiar circumstances described in this problem apply to any two points on a rigid rotating object: although they are at rest relative to each other, from the point of view of an external observer they have completely different velocities. If they were not riding alongside each other, then their velocities would not even be parallel.

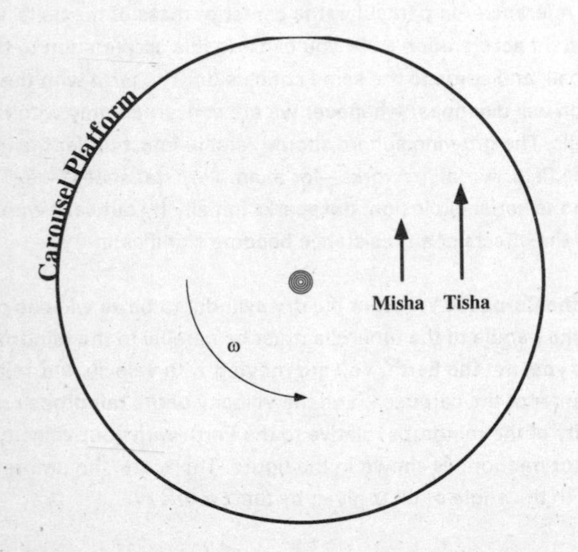


**7. At the Carnival Again—Misha and Masha:** Consider Masha's reference frame. The rest of the world revolves about the center of her carousel with velocity $\omega$. Misha's motion (in this reference frame) must then be comprised of two components: the motion about the center $O_1$ of Masha's carousel and his motion relative to the center $O_2$ of his own. As shown in the figure, when they pass each other, the two velocity vectors will point in exactly opposite directions. Now Misha is 3 meters from the center of Masha's carousel, so the magnitude of the first component will be $3\omega$, resulting in the vector labeled $\mathbf{u}'$. Remember that this is a *negative* velocity from Masha's point of view. In the figure, she faces in the direction of the arrow. The second component, $\mathbf{u}''$, Misha's motion relative to his own carousel, is just $2\omega$. This velocity is added to $\mathbf{u}'$. Therefore, the two components do not cancel, and in fact there is a net negative linear velocity of $-\omega$, so the relative velocity has such a direction that, from Masha's viewpoint, Misha is going backward. Clearly she appears to be overtaking and passing him.

However, it is clear that this situation is symmetrical, and if instead we begin with *Misha's* reference frame, it will appear to him as if he is passing her! You may

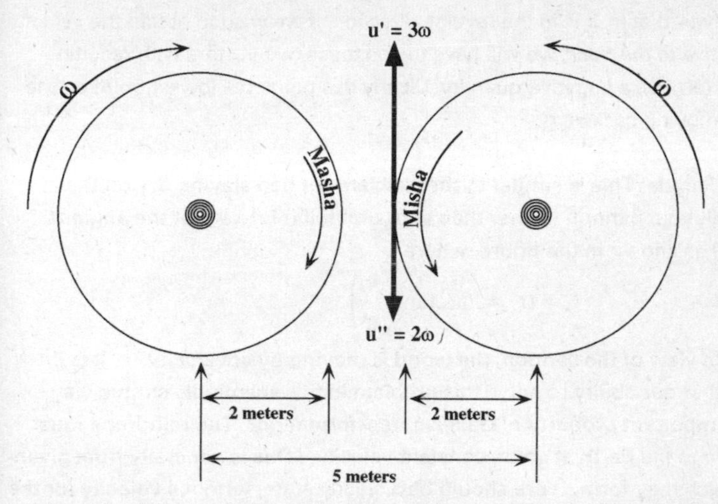


wish, with a friend, to prove this to yourself by experiment. The world of noninertial coordinates, particularly rotating frames of reference, is strange indeed, but it is quite consistent with cultural relativism and political correctness. It appears to us that many modern politicians live in rotating coordinate systems.

**8. Old Man Mazay and the Streetcar:** Old Man Mazay is looking at the lowest part of the streetcar, the lower edges of the wheels. Consider the figure, which is a sketch of one of the wheels of the streetcar, on the track. The flange extends over the track, below the point of contact of the wheel with the track. The wheel rotates with some angular velocity $\omega$, and the streetcar moves forward relative to the track with linear velocity $V_o = \omega r$, where $r$ is the radius of the wheel. The velocity of the lowest edge of the flange *relative to the streecar* is $V_r = \omega R$, where $R$ is the radius of the flange, and this

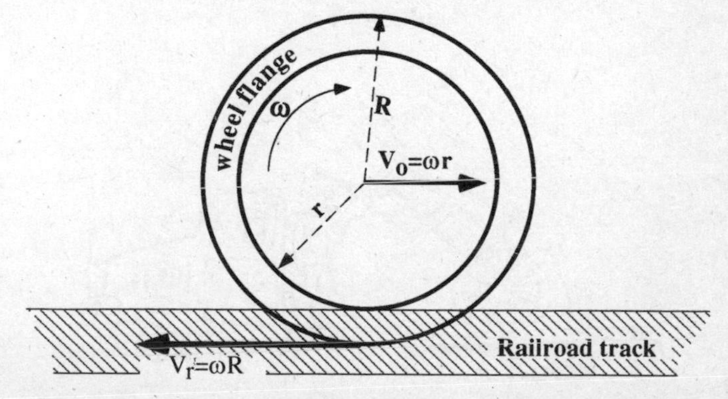

velocity is negative—that is, it is in the reverse direction. If we wish to obtain the velocity of this point *relative to the track,* we will have to add these two vectors with resulting velocity equal to $\omega(r\text{-}R)$, a negative quantity. Clearly this point, the lowest point on the flange, is in fact moving backward.

**9. The May Day Parade:** This is similar to the problem we had staying dry on the carousel! Now tilt your cannon (rather than your umbrella) forward at the angle $\alpha$ from the vertical as shown in the figure, where

$$\alpha = \arctan\left(\frac{v}{u}\right)$$

From the point of view of the cannon, the world is moving by at velocity $-\mathbf{v}$. It is important to note that our ability to solve this problem in this seemingly simple way depends on an important property of Galilean transformations. The raindrops must be falling relative to the Earth at some constant velocity. (This is generally true; given the height at which they form, there should be a steady state, terminal velocity for the raindrops.) If the velocity vector is indeed constant, the uniform, rectilinear property of motion of the raindrop will also occur with reference to any other frame that moves at constant velocity relative to the Earth. *Therefore, the trajectories of the raindrops must be straight lines from the viewpoint of an observer (you!) sitting in the truck.* It should be possible to set the cannon barrel at an angle such that the raindrops would pass completely through, never hitting the interior walls. The barrel must be set up so that its axis is parallel to the velocity $\mathbf{u'}$ of the raindrops *in the frame of reference fixed to the traveling cannon.* If $\mathbf{u}$ and $\mathbf{v}$ are the velocities of the rain and the cannon relative to the Earth, the velocity $\mathbf{u'}$ of the drop relative to the cannon is given by the Galilean transformation

$$\mathbf{u} = \mathbf{v} + \mathbf{u'}$$

or

$$\mathbf{u'} = \mathbf{u} - \mathbf{v}.$$

We can see from the figure that the cannon should be tilted forward and that the angle $\alpha$ can be easily found from the picture. The angle $\alpha$ is determined by the difference, not the sum, of **u** and **v**. If the cannon were not moving, $\alpha = 0$, that is, the barrel is vertical, an intuitive solution. You have probably concluded by now that the breach of the cannon should get quite wet in any case, which will do the cannon no good at all. However, your sergeant did not mention the breach, only the barrel! (Do you think this will help when they ship you off to the gulag?) (N.B.: Only one of us [R.R.] thinks this last remark is at all funny.)

**10. With a Roll of Drums, the Galilean Frame of Reference:** The figure given with the question shows the drum rotating between the plates with angular velocity $\omega$. We wish to determine that velocity. It is tempting to use the axis of the drum as the frame of reference, but it is more instructive and elegant to use the lower plate. Look now at the figure here.

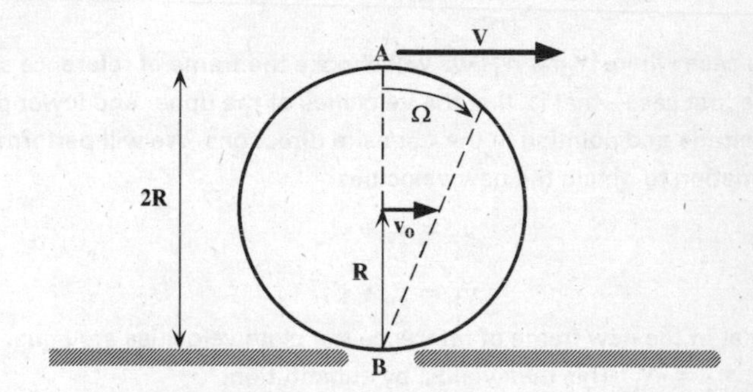


As it shows, the instantaneous center of rotation of the cylinder in this frame of reference is at the point of contact with the lower plate (B in this figure). If the drum rotates *about this center* at angular velocity $\Omega$, the speed $v_o$ of the axis of the drum must be $R\Omega$, and the velocity of the point A on top, in contact with the upper plate, is $2R\Omega$. We also know the relationship in this frame between the speed $v_o$ of the axis of the drum and the rotational velocity $\omega$ of the drum about its own axis: $v_o = \omega R$. Therefore, $\omega = \Omega$. For the special case $\mathbf{v}_U = -\mathbf{v}_L$, the velocity of point A is $2\ v_L$ and $v_o = v_L$. The rotation frequency (revolutions per second) is just $\omega/2\pi$ and is therefore equal to $v_L/2\pi R$. For the general case, the velocity of point A in this reference frame will be $\mathbf{v}_U$-$\mathbf{v}_L$ and then the rotation frequency will be

$$f = \omega/2\pi = |\mathbf{v}_U - \mathbf{v}_L|/4\pi R.$$

To use the axis of the drum as a reference, look at this second figure. For the first case $\mathbf{v}_U = -\mathbf{v}_L$ and $|\mathbf{v}_U| = |\mathbf{v}_L| = v$, the motion of the plates is symmetrical, and the axis of the drum remains at rest, rotating about its axis. The linear velocity

$v = v_U = v_L$ of the points of contact A and B is related to the angular velocity of the drum $\omega$ by the relation $v = \omega R$. One revolution of the drum corresponds to a rotation of $2\pi$ radians, and therefore the rotation frequency (revolutions per second) $f$ is given by

$$f = \omega / 2\pi = v_U / 2\pi R = v_L / 2\pi R.$$

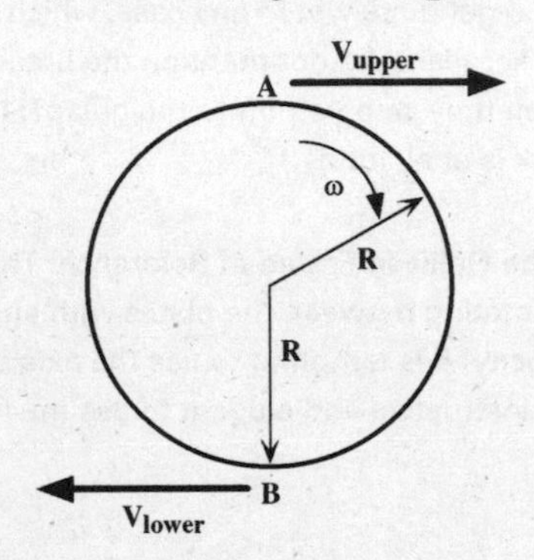


In the general case where $|\mathbf{v}_U| \neq |\mathbf{v}_L|$ we will choose the frame of reference such that it reduces to the first case—that is, that the velocities of the upper and lower plates are equal in magnitude and pointing in the opposite directions. We will perform a Galilean transformation to obtain the new velocities:

$$\mathbf{v}_L = \mathbf{v}_0 + \mathbf{v}'_L$$

$$\mathbf{v}_U = \mathbf{v}_0 + \mathbf{v}'_U$$

and require that in the new frame of reference the plate velocities are equal and opposite, that is, $\mathbf{v}'_L = -\mathbf{v}_U$. This then yields, by substitution,

$$\mathbf{v}_0 = \frac{\mathbf{v}_U + \mathbf{v}_L}{2} \quad \text{and} \quad \mathbf{v}'_U = \frac{(\mathbf{v}_U - \mathbf{v}_L)}{2}$$

The number of rotations is independent of the lateral motion of the frame. We can now use these equations to obtain the rotation frequency again:

$$\omega = \frac{\mathbf{v}'_U}{R} = \frac{|\mathbf{v}_U - \mathbf{v}_L|}{2R}$$

and therefore

$$f = \frac{\omega}{2\pi} = \frac{|\mathbf{v}_U - \mathbf{v}_L|}{4\pi R}$$

**11. Boris and the Wet Basketball—Reference Frames and Fluxes:** For once, Boris is correct, but so is Tisha. First consider the ball at rest. If we observe this ball from above, the target area that can be struck by the falling rain, the "circle of wetness" in the first figure, is a circle with a radius equal to the radius $R$ of the ball and area equal to $\pi R^2$. If we define $N$ to be the number of raindrops per unit volume, then the number of raindrops that strike the stationary ball per unit time is:

$$I_s = Nu(\pi R^2)$$

where $u$ is the speed of a raindrop.

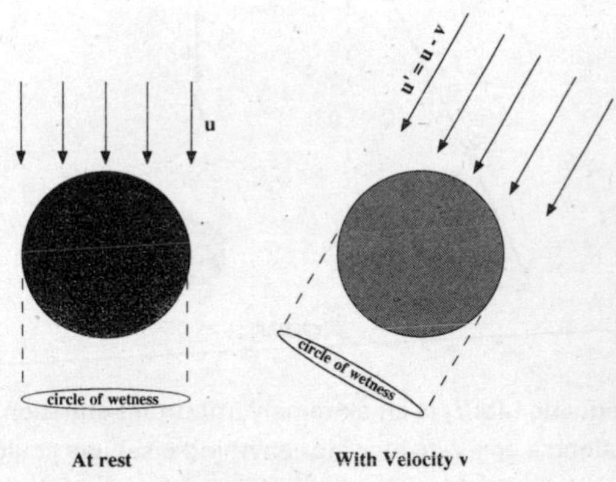


FIGURE 1

Now consider the ball sliding along with speed v. The raindrop density $N$ remains unchanged, but the falling rain no longer has the same velocity relative to the ball. Making a Galilean transformation, we choose a reference frame relative to the ball and, as shown in the second figure, the new velocity of the rain is $\mathbf{u'} = \mathbf{u} - \mathbf{v}$. Look back again at the first figure: the raindrops strike the ball at a different angle, but because of the ball's symmetry, the target area is still the same. The only difference between the sliding ball and the ball at rest is the relative speed at which the raindrops are striking the balls. Using the Pythagorean Theorem, we can see from the second figure that the speed of the raindrops striking the moving ball is

$$u' = \sqrt{u^2 + v^2}$$

and when $u'$ is used instead of $u$ in the equation for $I_s$, $I_s$ will be larger: there will be more raindrops striking the ball *per unit time.* So Boris is correct in his statement that the basketball will get wet faster if it moves faster. On the other hand, Tisha's statement is concerned not with the rate at which they will get wet but with how wet they will get all together. If, for instance, Boris doubles his (and the ball's) speed v, he can cut by half the time it takes to get home, and because

$$u' = \sqrt{u^2 + v^2}$$

this will not double the relative raindrop velocity or speed $u'$, and therefore $I_S$ will not double. Therefore, Tisha is also correct: by increasing his speed, Boris will not be as wet when he gets home.

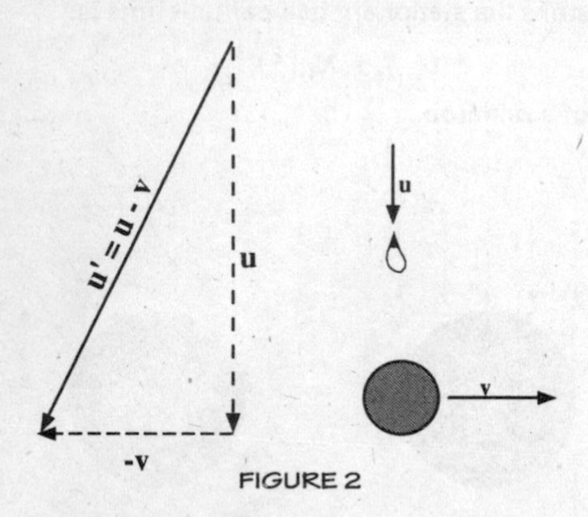


FIGURE 2

The equation for $I_s$ is an extremely important equation. If, for instance, Boris were pushing along a concrete block (or anything else), we could calculate the rate at which raindrops strike the block by determining the area of the horizontal cross-section area of the block and multiplying by $Nu'$. Such a quantity, the product of the average density ($N$) and the appropriate relative speed ($u'$), is called a *flux*. The dimensions of the flux are quantity (number of raindrops in this case) per unit area per unit time. The quantity can be number of particles or anything carried by the particles—for instance, energy, mass, or electrical charge. Any particle (including photons) can be used. If the quantity were, for instance, electrical charge, the flux would be the electrical current density. The idea of flux is central to all of science and engineering.

**12. Chicken Feed:** The bucket will fill at the same rate whether Boris walks or runs. The rain is falling straight down because there is no wind. If we consider some horizontal area, and if the rain is falling uniformly and at constant velocity, equal volumes of rain must pass through equal areas during equal times. No matter how fast or slow Boris moves, the area of the mouth of the bucket does not change. So the bucket will fill at the same rate unless Boris goes so fast that (a) Lorentz contractions occur, or (b) the bucket flies off the back of the wheelbarrow. What does happen, however, as Boris goes faster, is that more of the rain will wash down the back wall of the bucket, rather than falling straight into the bucket. This does not, however, change the total amount of rainwater accumulating in the bucket. On the other hand, this means that running is helpful. If Boris covers the same distance in less time, the bucket will definitely have less water in it.

**13. Fluxes and Conservation Laws (or It Always Helps to Run in the Rain):** There is no inconsistency. Review the solutions to the previous two problems. Because the ball is spherical, the "circle of wetness" has a constant area that does not depend on the velocity **v** of the ball. As the speed v increases, the relative velocity $\boldsymbol{u}'$ changes and the relative speed $u'$ increases, so the flux $I_S = Nu'(\pi R^2)$ increases. Now consider what the "circle of wetness" for Boris's bucket looks like, using the traveling bucket as a reference frame but *from the point of view of the raindrops outside the bucket.* Look at the second figure in the solution to problem #11. As the speed v increases, the angle $\alpha$ to the vertical of the relative velocity $\boldsymbol{u}'$ will increase—that is, the rain will be falling more horizontally. The mouth of the bucket now will not look like a circle but rather like an ellipse, as shown here in the figure.

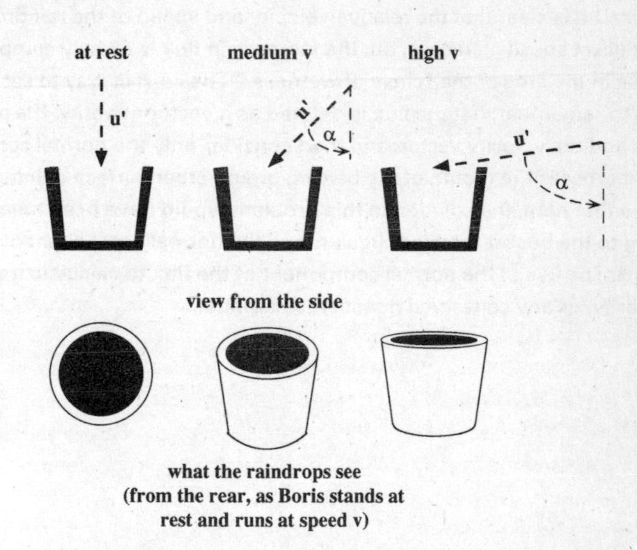


what the raindrops see
(from the rear, as Boris stands at
rest and runs at speed v)

If the rain were falling completely in the horizontal direction, the ellipse would close to a line and all the rain would hit the side of the bucket. This ellipse is our "circle of wetness." Since the area of an ellipse is just $\pi ab$ (where $a$ and $b$ are the semimajor and semiminor axes, or the two "radii" of the ellipse), the area can be calculated if the angle $\alpha$ is known. For the bucket we would have $a = R$, where $R$ is the radius of the mouth of the bucket, and $b = R\cos\alpha$. From the solution to problem #11, we can see that

$$\cos\alpha = \frac{u}{\sqrt{u^2 + \text{v}^2}}$$

and therefore that the area of the "circle of wetness" is

$$\frac{\pi R^2 u}{\sqrt{u^2 + v^2}}$$

We can obtain the total rate of input into the bucket by multiplying this area by the magnitude of the flux $Nu'$. Since

$$u' = \sqrt{u^2 + v^2}$$

our answer is

$$I_s = Nu\pi R^2$$

which depends only on the speed of the rainfall $u$ and *not* on the velocity or speed of the bucket! It is clear that the relative velocity and speed of the raindrops will increase as the bucket speed increases, but the increase in flux is exactly compensated by the decrease in the area of the "circle of wetness." The correct way to set up such problems is to remember that the flux is defined as a vector quantity, the product of the density and the velocity vector, and then consider only the normal component of the flux as it crosses the mouth of the bucket, or any other surface of interest to us. Had we done that here, the solution to this problem would have been much simpler: the opening to the bucket remains circular, and the normal component of the flux is $Nu'\cos\alpha$. The use of the normal component of the flux to calculate transfer of material, energy, or any conserved quantity is essential.

# Solutions to Chapter 5

## ON THE ROAD

**1. The Cossack and the Goat:** First, a graphical solution. Given the velocity vectors $\boldsymbol{v}_1$ and $\boldsymbol{v}_2$, the problem is that both the Cossack and the goat are moving. We fix the frame of reference to the Cossack (a typical Russian failing). In such a frame, the Cossack remains at rest and the velocity of the goat is its relative velocity $\boldsymbol{v}_{21} = \boldsymbol{v}_2 - \boldsymbol{v}_1$, as shown in the illustration.

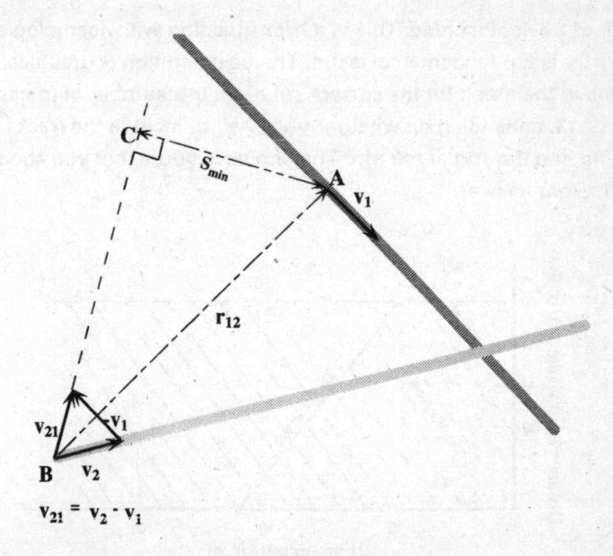


Now the goat travels, in this frame of reference, along the line BC, in the direction of the vector $\boldsymbol{v}_{21}$. The minimum distance $S_{min}$ can be obtained by dropping the normal AC from point A to the straight line BC. Then

$$S_{min} = |AC| = |AB| \sin \angle CBA.$$

As you can see from the figure, the Cossack doesn't get very close to the goat, which is probably healthier for the Cossack anyway. This problem is very familiar to radar operators, but in reverse. The radar screen shows the relative motions of all objects in the vicinity; for instance, for a boat, the screen will show other boats, buoys, and any nearby land. The operator must determine the absolute motions from the relative motions (a buoy, for instance, should definitely have no absolute velocity!), as well as the distances of closest approach.

There is also an analytical solution that is coordinate-free. We will outline this method, but not go into detail. Begin with the position vectors

$$\vec{r}_1(t) = \vec{r}_{10} + \vec{v}_1 t$$

$$\vec{r}_2(t) = \vec{r}_{20} + \vec{v}_2 t$$

where $\boldsymbol{r}_{10}$ and $\boldsymbol{r}_{20}$ are the initial position vectors of Cossack and goat, and minimize the magnitude of the vector difference (it is more convenient to minimize the square of the difference) between the two vectors. A most challenging task now would be to show that both answers are identical!

**2. A Triumph of Soviet Planning:** This is a trick question with cosmological implications. Symmetry is the fundamental issue. The best solution is graphical.

Look at the graph for the correct solution: the number of trucks passed will be either 11 or 13, depending on whether you wish to include the trucks passed at the very beginning and the end of the trip. The choice is yours, but you should not forget to specify it in your answer.

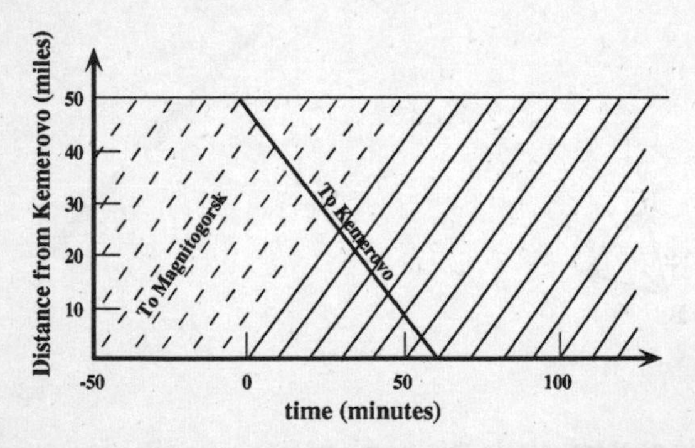


Here is the trick: we said that trucks left Kemerovo *every 10 minutes,* and that is *exactly* what we meant! No initial time for the problem is specified, and the starting point for the truck can be shifted (translated) along the time axis by any integral multiple of 10 minutes. In other words, we have *translational symmetry* along the time axis—the trucks start right after the Big Bang and continue until Judgment Day. Incidentally, if you calculate the frequency of the meetings as a function of the trucks' speeds, you will have derived the Doppler Effect, the explanation of the "red shift" in radiation from faraway stars. In any case, we should point out that the idea that the trucks proceed throughout all of eternity is central to the Soviet view of human affairs.

**3. Backing into the Answer:** First consider pulling out of the space, as shown in the figure. Once we have successfully pulled out, we can play back the whole scene in reverse. The front wheels are turned, and the rear wheels remain aligned, sharing the same axis (the rear axle). We draw a straight line through the rear axle, and the center of rotation $O$ will occur somewhere on this line: faraway if the front wheels are turned only slightly, closer in for a tighter turn. This arrangement applies to both cases, pulling in and pulling out. The center of rotation is closer to the back of the car and therefore, in pulling out, the front of the car will swing farther out, and it will be easier to avoid the car in front. The time-reversed version of this will have us backing in!

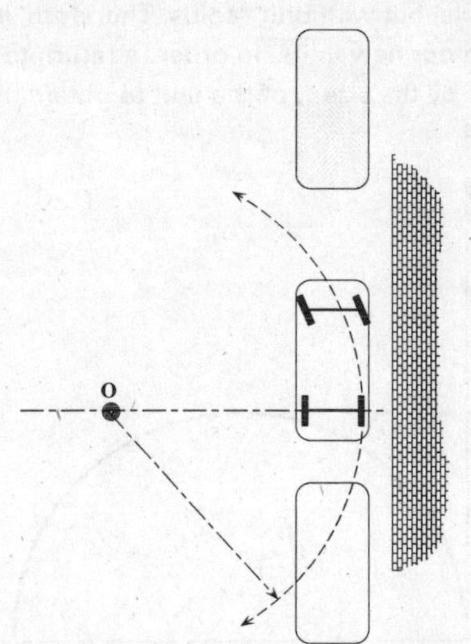


**4. Another Driving Problem:** Look at the solution to the previous problem, but now consider the front wheels. The center of rotation of the car lies somewhere on a line drawn through the rear axle, far away from the car if the car is turning gently, closer in if the turn is more extreme. If the front tires are exactly parallel to each other, they cannot have the same center of rotation! To have the same center of rotation (and avoid excessive forces on the front end and tires), they must turn by different angles. However, if they are set permanently at different angles, there will be scuffing and wear when the car is moving in a straight line or turning at a different rate. Suspension design has allowed for this problem since the 1920s, but only approximately. This is only one problem faced by the designer, who has to worry about the dynamic behavior of the front end, which we will not discuss here. For a number of reasons, the front wheels are anything but parallel. The perfect suspension, which would turn the wheels at exactly the proper angles, has yet to be designed.

**5. How Far Did That Car Go? (A Trip Through the Wrong Dimension):** The custodian is correct. He noticed that each of the mathematicians' answers did not make sense because it did not have the dimension of length. Clearly, the answer should be given by the area under the curve, but it must have the dimension of length (miles or kilometers or whatever). In order to circumvent the difficulty, we can take new units for our variables: $T = t_o/2$ as a new unit of time and $v_o$ as a new unit of velocity. The unit of length then will be $L = Tv_o$. If we now plot the dimensionless velocity $v/v_o$ versus dimensionless time $t/T$, we obtain the graph shown in the figure below. It is still a semicircle, but with unit radius. Therefore, it has the area of $\pi/2$, which is dimensionless in our new units. In order to return to the usual units, we must multiply the answer by the size $L$ of the unit to obtain the distance in question as $s = \pi L/2 = \pi v_o t_o/4$.

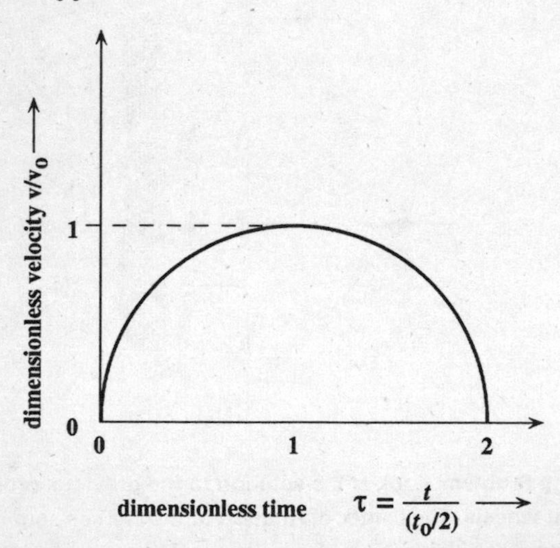


This silly example is a good demonstration of how the use of dimensionless variables can keep you out of trouble, and why they are so popular in all fields of physics. Hydrodynamics, aerodynamics, and heat transfer, for instance, would be much more difficult than they already are without the use of dimensionless variables.

**6. Where Did That Car Go Now?:** Graph B must be incorrect. You may remember why the graph of a function never has two points in the same vertical line. In any case, in graph A the vertical axis represents the speed, the absolute value of the velocity, which has to be positive. In graph B the vertical axis represents only $v_x$, the component of the velocity along the x-axis, which can be positive or negative depending on the direction of motion. However, graph B shows that $v_x$ is always positive, that is, that the position on the x-axis should always be increased. How, then, can the car manage to move from $x_2$ back to $x_1$? How can it decrease its x-coordinate when it has only positive velocity in that direction? It cannot. Graph B shows contradictory information and so cannot be correct.

Since graph A shows the magnitude of the velocity versus $x$, the argument that proved graph B to be incorrect cannot be used again. Although the *magnitude* of the velocity may be positive throughout, the x-component of the velocity may be negative. (After all, the magnitude is defined as the length of the vector, which cannot be negative.) It is true that the car may not have two different velocities at the same point at the same time, but graph A makes no such statement. Graph A states that the car may have different velocities at points having the same x-coordinate. Such a statement may be made for any spatial axis (x, y, . . .) but not for the time axis. Clearly, it is impossible to have different velocities at points having the same time coordinate. This is an important property of the time-space continuum.

We will leave it up to you to graph the actual motion of the car as described by graph A. You might also note that our race car designers have not, as yet, learned to use dimensionless variables.

Both graphs possess an interesting feature: they have points for which two or more values of $x$ correspond to one value of $y$. Similarly, we can see points for which two or more values of $y$ correspond to one value of $x$. It means that the whole curve cannot be expressed as either $y = f(x)$ or as $x = F(y)$. The curve can be expressed either parametrically or as an equation $F(x,y) = 0$. There are mathematicians who despair of the mathematical proficiency of physicists, saying "physicists believe that every variable is a function of all the other variables." The physicists generally defend themselves by saying that economists, biologists, and other applied scientists are certain that every variable is a *linear* function of all the other variables. The latter respond by saying that the rest of the world does not even understand the accusations.

A custodian with sufficient education and intelligence to spot such problems was not uncommon in Soviet Russia. Politically incorrect scientists usually found themselves in such situations, at the business end of a broom. Dr. Boris Altshuler, an associate of Dr. Andrey Sakharov and a brilliant theoretical physicist, was in a similar position for over a decade. One of us (Y.C.) was honored by his company and friendship during that time.

**7. Winter Driving:** We will consider only the correctness of Petya's statement. There are other excellent reasons to drive carefully in winter, one of which will be discussed in the next problem. If the friction coefficient of the road were perfectly uniform and constant (a common condition in introductory physics!), Petya would be quite correct, provided he could react instantaneously or at least very quickly, as he claimed. However, in addition to the decrease in friction coefficient, there is a dramatic increase in fluctuation of the coefficient, from bare road to "black ice." In the language of the previous problem, the velocities may be identical but the positions are not, and therefore the braking distances are not identical.

The important thing is to include the fluctuations in the analysis. Sanding the road not only increases the friction coefficient; it decreases the fluctuations. Consider this mythical, expert statistical bowman: the average position of all his arrows was always exactly at the center of the target, but no individual arrow actually ever hit the target! In some circumstances, variability and fluctuation are most important!

**8. Winter Driving II:** It is absolutely necessary. The friction coefficient defines the maximum friction force between two bodies, and when you are skidding, or about to skid, the absolute value of the friction force between your tires and the road is equal or close to the maximum value. Suppose a body (your car, for instance) is being pushed or pulled along a horizontal surface (the road, for instance). If the applied force is less than the maximum friction force $F_{max}$, then the friction force simply opposes the applied force. The magnitude of the friction force will be equal to that of

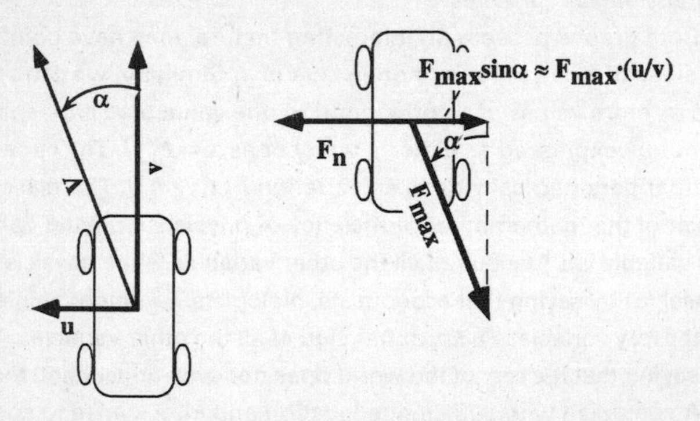


**Note: because u is not yet at the ultimate value, $F_n$ is larger than the lateral component of the friction. The skid will accelerate!**

the applied force, and the direction will be the opposite of the applied force. On the other hand, if the car is sliding, the direction of the friction force is no longer the opposite of the direction of the applied force, but instead is opposite to the direction of the velocity vector. Also, when sliding, the magnitude of the friction force cannot exceed the maximum value $F_{max}$.

Now consider the effect of a lateral force $F_n$ applied to a car that is sliding straight forward, as shown in the figure. There will be a lateral acceleration until the applied force is balanced by the friction force. For some skidding velocity $u$, the friction force will be directed at an angle $\alpha$ to the axis of the car. The lateral component of the friction force will be $F_{max}\sin\alpha$ or, if $\alpha$ is small, $F_{max}\cdot(u/\text{v})$. The ultimate skidding velocity will then be given by the equation

$$F_n = \left(\frac{u}{\text{v}}\right)F_{max}.$$

This equation resembles the law for fluid friction (lubrication) and shows us that if the forward velocity is large enough, even a small transverse force can cause sliding sideways with considerable speed.

Before this balance is achieved, there will be a lateral acceleration that obeys Newton's Second Law, that is

$$ma = m\dot{u} = F_n - \left(\frac{u}{\text{v}}\right)F_{max}.$$

This equation is instructive because it looks like the equation of motion with fluid friction. The sliding in the forward direction creates "virtual" lubrication for the transverse sliding, that is, it reduces the second term on the right side. Driving slowly will increase this "friction" term and reduce lateral motion. Much safer!

**9. Another Triumph of Central Planning:** The key to a rigorous, absolute proof is to plot the odometer distance along each road as a single coordinate.

A point on this graph will represent the position of *both* vehicles. Now one of the trajectories on the graph will correspond to the two cars proceeding from Serpukhov to Chekhov on the old and new roads. The other will correspond to the two trucks, one from Serpukhov to Chekhov and the other vice versa, on the old and new roads. At the point of intersection, the arrangement of the trucks (their positions) is exactly the same as a previous configuration of the cars. Now we know this will not work, for there must be no more than 20 meters between the trucks at that point, because the cars were connected by a 20-meter telephone wire at that moment. The

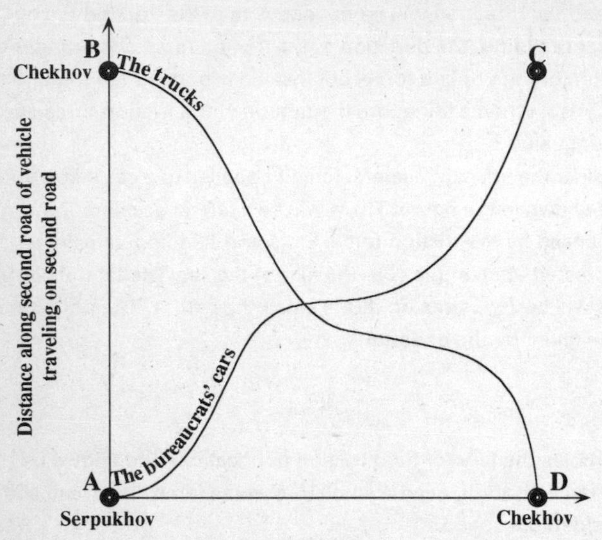


Point A: both vehicles are in Serpukhov
Points B and D: one vehicle is in Serpukhov and the other is in Chekhov
Point C: both vehicles are in Chekhov

graph makes it clear that such a situation is inevitable. Both truck drivers better have their seat belts fastened! This method of plotting may strike you as unusual: a point on the graph represents the position of *two* vehicles. However, this sort of thing, called "state space," is used in the theory of dynamical systems and statistical mechanics as a very powerful tool to represent generally the states of various complex systems.

# Solutions to Chapter 6

## SOURCES AND SINKS

**1. A Plumbing Problem:** Consider the stream of water at any two heights. In fact, cut the stream with two horizontal planes, and measure the area of the stream at these two points. Because the shape of the stream does not change with time, equal volumes of water per unit time must pass through each of the two planes. If they did not, then there would be either an increase or a decrease of the water between the planes, and the shape of the stream would be changing with time. (To conserve mass, the increase or decrease would be equal to the difference in the flow across the two planes. Also, conservation of mass and conservation of volume amount to the same thing because the density of the water is constant in this case.) Since we know that the shape remains constant, then at any height or distance from the faucet, the flow rate must be a constant. However, the velocity of the water must change as it descends, due to the acceleration of gravity. To keep the total flow rate constant, then, the cross section of the stream must decrease. Suppose the faucet were dripping instead of streaming. Consider a single drop, accelerating as it falls. The distance between successive drops increases as they descend. If one imagines the stream as a collection of drops, one can see why the stream gets thinner as it leaves the tap.

This problem illustrates one form of the principal of conservation and continuity: in the steady state, for the flow of any quantity that is conserved, the rate of mass flow—that is, the velocity times the cross-sectional area—always equals a constant. Therefore, if one increases the velocity, the area shrinks accordingly. What do we mean by "steady state"? We mean that the properties of the system—for instance, the shape of the water stream—do not change with time. If the rate of mass flow is not a constant, then changes will occur: matter will pile up in (or leave) the vicinity, and the shape of the water column (or something else) will change with time.

The results of the neglect of this principle can be readily observed in the management of traffic flow in the Boston area, where the speed limit is reduced whenever the road narrows. Such regulations make the achievement of steady-state flow impossible, and the roads of eastern Massachusetts are notorious for the resulting traffic delays. (Perhaps the highways of Massachusetts should be turned over to the water departments.) In Germany, the usual practice is to place a *"Schnell!"* sign wherever the road approaches a narrow spot.

**2. More Practical Advice from Your Plumber (How to Measure Gravity with a Ruler and a Bucket):** Yes, you can! Look at the previous problem. The stream becomes thinner as it descends because the water is falling faster, as any falling object would. The density of the water remains constant. Conservation of mass then dictates that the net mass flow rate should remain the same, and therefore the stream should thicken when it slows down and become thin when it speeds up. Here is what you should do: use the ruler to measure the diameter $D_o$ of the inside of the pipe or tap from which the water emerges; use the same ruler to measure the diameter $D$ of the water stream at some distance $y$ downstream from the tap; in fact, use the ruler to measure the distance $y$ as well, as shown in the figure.

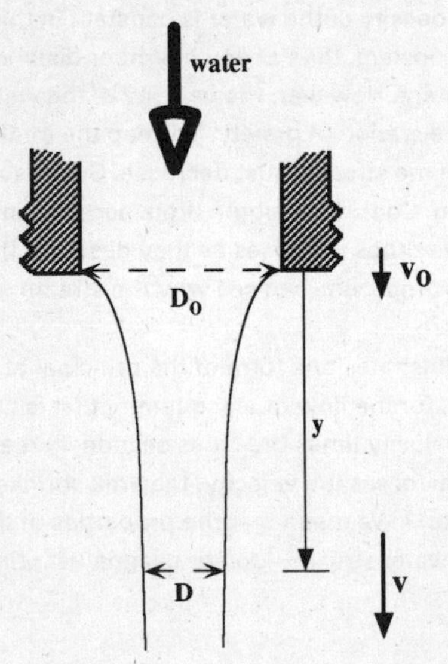

If the water leaves the tap with velocity $v_o$ we can write, using conservation of energy

$$\frac{1}{2}mv^2 = \frac{1}{2}mv_o^2 + mgy \qquad (1)$$

for any element or piece of the flow that has the mass $m$, where v is the downward velocity at position $y$. We can solve for v:

$$v = \sqrt{v_o^2 + 2gy} \qquad (2)$$

which is the usual equation for any falling object with intial velocity $v_o$.

The volumetric flow rate $Q$ can be thought of as the volume of a cylinder with the cross-section area of the water stream and length equal to $v\Delta t$, with $\Delta t$ equal to 1 second. $Q$ is therefore equal to the product of the cross-sectional area and the velocity, and must be the same at all points in order to conserve mass, so

$$Q = \frac{\pi}{4}D^2v = \frac{\pi}{4}D_o^2v_o \text{ or } D^2v = D_o^2v_o \qquad (3)$$

We can use equation 2 to substitute for the velocity in equation 3 and get

$$D^2v_o^2 = D^4(v_o^2 + 2gy) \qquad (4)$$

which can be solved for $v_o$, the velocity with which the water is leaving the tap.

$$v_o = D^2\sqrt{\frac{2gy}{D_o^4 - D^4}} \qquad (5)$$

The right-hand side of this equation consists of quantities that can be measured with the ruler and the gravitational acceleration $g$. Finally, the velocity can be substituted into equation 3 to obtain

$$Q = \left(\frac{\pi}{4}\right)D_o^2D^2\sqrt{\frac{2gy}{D_o^4 - D^4}} \qquad (6)$$

the volumetric flow rate.

It is clear that if flow rates can be measured and we still have our ruler around, we could solve equation 6 for $g$. A 5-gallon bucket and a wristwatch will suffice to obtain two or three significant figures for the flow rate. Try it!

**3. "You Have 20 Seconds to Make Your Calculation Then I Strike!":** We need consider only the forces in the vertical direction. The simplest solution uses Newton's Second Law in its most basic form, that is, that the sum of the forces on any body must equal the time rate of change of the momentum. However, the momentum consists of two factors: mass and velocity. In most familiar cases, the momentum changes by changing the velocity, that is, by acceleration. It is clear from the definition of the momentum as $m\mathbf{v}$ that changing the mass involved will also change the momentum. The garden hose is again a good example of this, as unbalanced forces can be experienced with no accelerations. In this case, the force comes from the fact that the masses involved do change; the mass at rest decreases and the mass with speed v increases. Now write down Newton's Second Law:

$$\frac{d\vec{p}}{dt} = \sum(\vec{forces})$$

Note that the cobra's head does *not* accelerate. Although the head of the cobra rises with constant velocity, the momentum changes in the vertical direction because the vertical length of the cobra (and therefore the mass that is moving vertically) is increasing with time! If the vertical length of the cobra is $h$, the mass of the cobra that is in the vertical segment is $h(M/L)$, and the component of the momentum in the vertical direction is written as

$$p_y = h\left(\frac{M}{L}\right)\mathrm{v}$$

then the rate of increase of this component of the momentum is simply a time derivative of the above expression, which yields

$$\frac{dp_y}{dt} = \left(\frac{M}{L}\right)\mathrm{v}\,\frac{dh}{dt} = \frac{M}{L}\,\mathrm{v}^2$$

According to the Second Law, the rate of increase of momentum must be equal to the sum of the gravitational force $Mg$ and the reaction $N$ of the ground on the cobra's tail. Since $Mg$ and $N$ have opposite directions, we obtain

$$\frac{M\mathrm{v}^2}{L} = N - Mg.$$

Thus, we finally find the absolute value of the reaction force $N$ as

$$f = M\left(\frac{\mathrm{v}^2}{L} + g\right)$$

According to Newton's Third Law, this should be equal to the force exerted *on the snake* by the ground. The term $\mathrm{v}^2/L$ resembles the usual expression for centripetal acceleration or centrifugal force, but this is fortuitous.

The problem can be solved this way because the Second Law can be interpreted as a balancing relationship, where forces are described in terms of credits and debits of momentum. This balancing relation boils down to the law of conservation of momentum when the sum of the external forces is 0. The fact that changes in mass can be involved then follows naturally. Engineering problems involving rocketry and space vehicles must take this into account.

**4. Conserving Momentum (and Energy!) in St. Petersburg:** Oblomovetz, the sleeper, goes much farther. Consider his situation first. The only change is that mass (in the form of falling snow) is being added to that of this lazy sweeper and his trolley. There are no interactions with horizontal outside forces and therefore, according to Newton's Second Law, the total momentum in horizontal direction must be conserved. If $m_o$ is the original mass of the trolley with Oblomovetz, then the mass at any time $t$ is:

$$m(t) = m_o + \mu t.$$

The original momentum was $m_o v_o$. Since the total momentum of this system must remain constant with time, $m_o v_o = m(t)v(t)$. We can now solve for the velocity of Oblomovetz and his trolley as a function of time. In doing so, we obtain:

$$v(t) = \frac{v_o}{1 + \frac{\mu t}{m_o}}$$

Now examine the situation of the workaholic, Stakhanovetz. When the snowflake lands on his trolley, it acquires the velocity of the trolley and, therefore, some momentum. Because Stakhanovetz is sweeping the snow off the side of the trolley as soon as it lands, he is actually "throwing away" some of the momentum of his system. Therefore, the workaholic's trolley system is no longer isolated from its surroundings, and the law of conservation of momentum can no longer be applied. Let us determine what effect this work has on the velocity as a function of time.

Since no snow is allowed to accumulate on the trolley, the mass is constant and the change of momentum of Stakhanovetz and his trolley is going to be based solely on the change in velocity:

$$\frac{dp}{dt} = m_o \frac{dv}{dt}$$

The mass of the snow descending during time interval $dt$ is $\mu dt$. Since each snowflake acquires velocity v, the momentum acquired by the snow when it hits the trolley is $(\mu dt)v$. This momentum is lost when Stakhanovetz sweeps the snow off the trolley, that is

$$dp = -v\mu dt.$$

Combining these two equations, we obtain the equation:

$$\frac{dv}{dt} = -\left(\frac{\mu}{m_o}\right)v$$

What function satisfies this equation? Whenever the derivative of a function is proportional to the function itself, it is the exponential function $e^{Ax}$:

$$d(e^{Ax})/dx = Ae^{Ax}.$$

This is the only function that has such a property. Look for solutions to the equation by substituting an exponential, remember the initial conditions ($v = v_o$ at $t = 0$), and you will find that the velocity of the workaholic sweeper as a function of time is

$$v = v_o e^{-\left(\frac{\mu}{m_o}\right)t}$$

Now in comparing the two equations, we can see that at any time the velocity of the lazy sweeper will be greater than the velocity of the workaholic. If you remember that the exponential function $e^{Ax}$ increases faster than any other function of $x$, you will see here that the function $e^{-Ax}$ will decrease faster than $1/(1 + Ax)$, so the velocity of the workaholic sweeper will decrease faster than that of the lazy sweeper. This is an important principle, and it is worthwhile to look up the rigorous proofs. The exponential function has greater convexity than any polynomial function; that is, the polynomial function will always have a highest-order nonzero derivative, whereas the exponential function never runs out of nonzero derivatives. We have plotted the two equations here, to demonstrate that statement in this particular case.

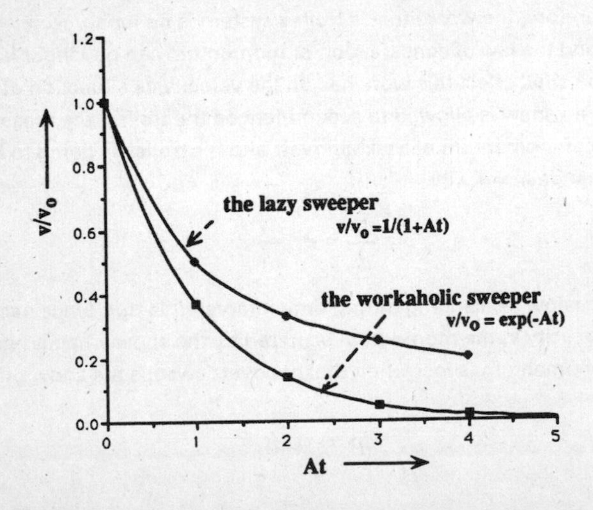

Our conclusion would change if the workaholic sweeper were not so naive. Consider, for instance, what would happen if he swept the snow in some direction other than perpendicular to the direction of his motion. At this point, you should be able to calculate the resulting behavior of his velocity. That, in the case we solved, the lazy sweeper travels farther than the workaholic is a general property of Socialist systems with the exception of leadership positions; no rigorous proof is available, but the empirical evidence is overwhelming.

**5. Mikhail's Merry-Go-Round:** We will show you two solutions, one based on angular momentum and one on the work done on the raindrops. The common and necessary concept in both analyses is the mass flux incident on the carousel. If there are $n$ raindrops per unit volume and the mass per droplet is $m$, then the mass density $\rho$ of the rain is given by $\rho = nm$. The raindrop flux, the number of raindrops landing on unit area per unit time is then $j = n\mathrm{v}$, and the mass flux, the mass of water landing on unit area per unit time, is $J = \rho\mathrm{v}$. Considering the carousel as a disk of radius $R$ and area $\pi R^2$, there will be $\pi R^2 n\mathrm{v}$ drops falling on the carousel per unit time. The total mass of water falling, per unit time, on the carousel is then $\pi R^2 \rho\mathrm{v}$. In each of the two cases, we will assume that the water falls off the carousel at the same rate that it is accumulating—that is, that the carousel is in the steady state, neither flooding nor drying up.

Now, for the first approach, consider that each raindrop landing on the carousel loses its vertical motion, and therefore loses the kinetic energy it had while falling. The rotating disk (our carousel) accelerates the drop as it slides outward to the edge, where it has the velocity $\mathrm{v} = \omega R$, where $\omega$ is the angular velocity of the carousel, while rotating. Just before it slides off the carousel, the drop has the kinetic energy

$$\frac{1}{2}m\mathrm{v}^2 = \frac{1}{2}m\omega^2 R^2$$

and this kinetic energy is lost when the drop slides off. Since the net number of departures is, we have assumed, equal to the number of arrivals, the rate at which the energy is lost is just the product of the raindrop kinetic energy and the rate of arrival, $\pi R^2 n\mathrm{v}$. The result is

$$\text{Energy loss rate} = \left(\frac{1}{2}m\omega^2 R^2\right)\pi R^2 n\mathrm{v}$$

$$= \frac{1}{2}\pi\rho\mathrm{v}\omega^2 R^4.$$

Mikhail's electric bill should reflect this increased energy use.

The second approach uses angular momentum considerations and is similar to the solution to problem #4 in this chapter. The drops land and acquire angular momentum from the carousel, which is lost when they fall off the outer edge. The loss must be compensated for by the motor driving the carousel. Consider a thin disk of water on the carousel that is about to be lost and renewed. The moment of inertia $dI$ of this disk, which has mass $dm$, is

$$\frac{R^2}{2}\, dm.$$

The angular momentum of the disk is $dM = \omega dI$, and the change in angular momentum must be compensated, if the carousel is to rotate at constant velocity, by torque $T$ from the motor:

$$\frac{dM}{dt} = T.$$

Since we have already determined in the first paragraph of the solution that the rate of arrival (and loss) is $\pi R^2 \rho v$, $dm = \pi R^2 \rho v dt$ in the time interval $dt$, and

$$T dt = dM = \omega dI = \omega \frac{R^2}{2} dm = \frac{1}{2} \pi \omega \rho v R^4 dt.$$

The torque necessary to drive the carousel in the rain is

$$T = \frac{1}{2} \pi \omega \rho v R^4.$$

Since the power needed is equal to $T\omega$, the final answer is

$$\text{Energy loss rate} = \omega T = \frac{1}{2} \pi \rho \omega^2 v R^4$$

again. Please note that we have assumed in this second solution that the disk of water rotates as a rigid, flat, uniform disk. In the first solution, no such assumptions were necessary. At this point, it is not too difficult to estimate the increase in Mikhail's electric bill, and we will leave this for you, but you should notice the dependence on angular velocity and also on the size of the carousel.

**6. Catching the Bus to Yakutsk (While Finding the Mach Cone):** Consider the bus at some time $t = t_1$. At that time it will be at position $x_1 = vt_1$ on the road. As the figure shows, to meet the bus *at that position* you will have to be within a circle of radius $ut_1$ centered at $x_1$. If you are outside the circle, it will be impossible for you to reach $x_1$ in the time $t_1$. The circle expands with time (or the position of the bus), and the envelope of all circles is a wedge (in three dimensions, this would be a cone enveloping an expanding series of spheres) with half-angle $\alpha = \arcsin(u/v)$, as shown. If you are outside the wedge or cone, it will be impossible for you to catch the bus, which in Siberia is a serious situation.

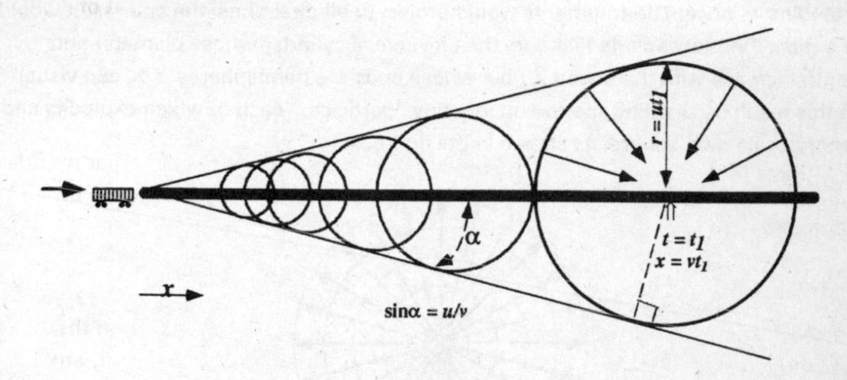


As for our "movie," it will show people running inward from the boundary of the cone at maximum velocity $u$ and disappearing into the bus; the cone shrinks so as to keep pace with the bus, and in fact appears to be pacing the bus. (The optimum direction to run is perpendicular to the boundary or wall of the cone. Can you demonstrate this?) If run in reverse, the cone now expands, and this can be represented by a series of expanding spheres or (in two dimensions) circles, for instance, continuous explosions that form spherical shock waves, or circular ripples on a liquid surface that continuously expand.

In fact, this is, in two dimensions, the bow wave of a boat. The bow wave can be represented as a sum of circular waves emanating from the bow as the boat moves. In order for this to happen, the boat speed must be higher than the wave speed, that is, $v \gg u$. A three-dimensional example of this phenomenon is the very strong shock wave caused by supersonic flight; this is referred to as the *Mach Cone*. The Mach Cone is fatal to unprotected observers at close (a few hundred feet) range. The electromagnetic radiation generated by fast electrons in solids (which move at velocities higher than the speed of light in that solid) is also emitted in a Mach Cone. It is called Cherenkov Radiation.

You may notice that nothing specific has been said about the optimum direction to run. Can you determine this? If you are standing on the boundary of the Mach Cone, is it better to run directly to the road in the perpendicular direction, or is it better to run perpendicular to the surface of the Mach Cone? We'll let you struggle with this one.

**7. An Explosive Topic: The Ice Cream (Mach) Cone:** First, ask what happens when a single, tiny particle explodes. It will explode in all directions, producing a spherical explosion. The radius of the sphere will increase with speed $u$. The sphere will expand with time, ultimately spreading out to infinity. Next, imagine a short rod, all of it exploding at once. The fragments would project in all directions: the end result would be a giant Tylenol capsule filling up the universe, a cylinder whose diameter and length increase with the speed $2u$ but whose ends are hemispheres. You can visualize this result by dividing the rod up into tiny "particles," each of which explodes and generates its own sphere, as shown in the figure.

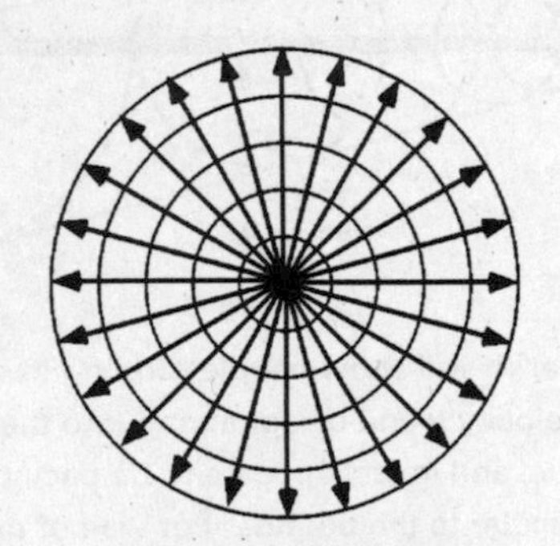

single particle, instantaneous

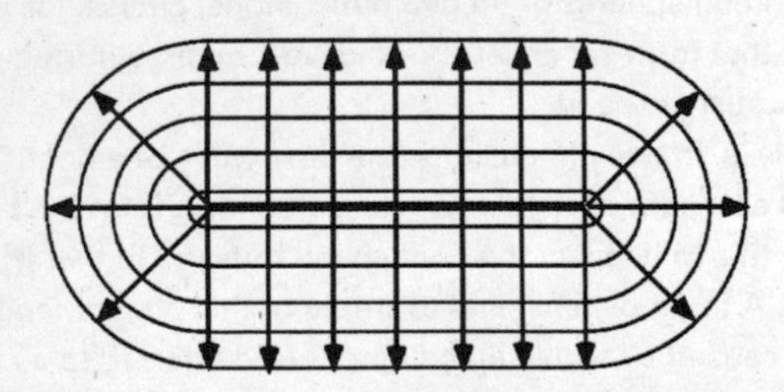

thin rod, instantaneous

Now for the semi-infinite rod. The strategy again is to consider the rod as a chain of tiny particles. Each particle generates a spherical explosion that spreads through the air with speed $u$. However, the particles do not explode at the same time. The explosion moves along the rod with speed v, which means that, if the end of the rod explodes at time $t = 0$, a particle $x$ meters from the end will explode at time $t = x/\mathrm{v}$. The result is a series of spheres, the largest sphere growing from the end and having radius $ut$, and as you proceed down the rod to later explosions, successively smaller spheres, until you arrive at the point where the explosion is just occurring. The successively shrinking spheres add up, as shown in this second figure, to a cone with altitude v$t$ with a hemisphere of radius $ut$ fitted perfectly to the base of the cone: a gigantic ice cream cone (with ice cream in it) growing to infinity.

This is another example of the Mach Cone.

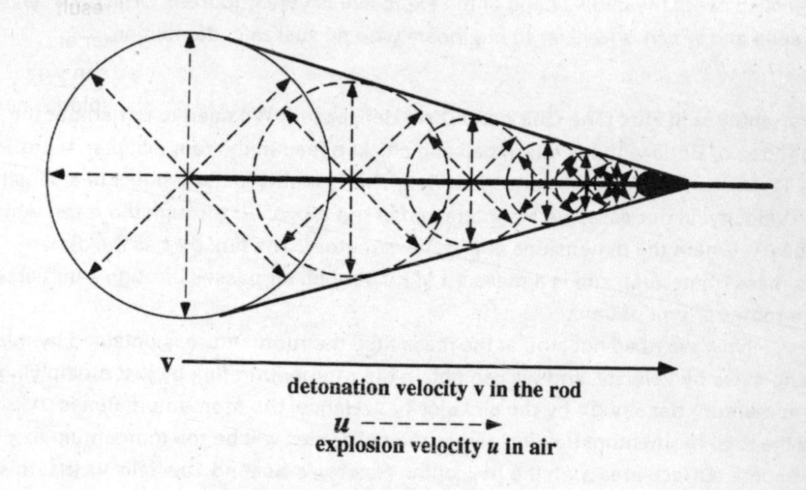


Semi-infinite rod, explosion begins at end

**8. More on High Explosives (and Ice Cream):** For the case v>$u$, the cone will grow, as in the previous problem, until it reaches the end of the rod, at time after initiation $t = L/\text{v}$, where $L$ is the length of the rod. As long as v$t < L$, the situation is identical to the semi-infinite rod. After the end of the rod is reached, the cone continues to grow outward, but the "point" of the cone now becomes a hemisphere, smaller than the one at the other end but growing at the same rate, with speed $u$. More technically, we have a truncated cone and hemispheres expanding outward with speed $u$ . The base and the truncated apex have hemispheres of radii $ut$ and $u(t\text{-}L/\text{v})$, respectively fitted onto the cone as shown in the second figure for the solution to the previous problem. The ice cream cone has become bloated—a common result of too much ice cream.

Now suppose that v is less than $u$, that is, the explosion velocity is low in the explosive rod compared to the speed of the explosion debris in space. In that case, the explosion path will simply be a sphere, centered on the end of the rod where the explosion began. Again, use the method demonstrated in the previous problem. The spheres generated by all other points on the rod will never catch up to the first sphere. In fact, this will occur (provided v<$u$) no matter what the shape of the explosive is! This is true even for the semi-infinite rod. Because the outer surface of the sphere is created only by the first "particle," the intensity of the explosion is spread out considerably over the interior of the sphere, diffusing the force of the explosion. In contrast, high explosives are characterized by high detonation velocities v, and in such cases the initial shape of the explosive is very important, which we have now seen and which is familiar to engineers who specialize in demolition.

**9. Everything Is in Flux (The Quick and Dirty Helicopter):** We need to remember the adventures of Boris with his basketball and chicken feed in the rain (chapter 4, problems 11–13), in particular the concept of *flux.* This quantity is the product of a density and a velocity. In our case, we are interested in the flux of air through the rotor, which will be $\rho v$, where the dimensions of $\rho$ is mass/meter$^3$. The flux $\rho v$ has the dimensions mass/(meter$^2$-s), and is a measure of how much air passes through a unit area of the rotor per unit of time.

Now we need not stop at the mass flux; the momentum is obtained by multiplying mass by velocity, and we can obtain the momentum flux by just multiplying the momentum density $\rho v$ by the air velocity v. Hence, the momentum flux is $\rho v^2$. Now the total momentum flowing across the helicopter will be the momentum flux times some surface area $S$ on the helicopter. Newton's Second Law tells us that this quantity, the rate of change of the momentum, is what keeps the helicopter up, by balancing the weight $M$ of the machine. Therefore

$$Mg = S\rho v^2 \quad (1)$$

We can use the fact that the mass $M$ is proportional to $L^3$ and write equation (1) for both the model and the real thing, dividing one by the other to get

$$\frac{v'^2}{v^2} = \frac{L'}{L} \quad (2)$$

The power $N$ should be given by the energy flux, which is the kinetic energy density

$$\frac{1}{2}\rho v^2$$

multiplied by the velocity, through some surface area $S$ of the helicopter. So we can say that the power needed to keep the model in the air is

$$N = \frac{1}{2}S\rho v^3 \quad (3)$$

As we mentioned in the hint, the area $S$ should be proportional to $L^2$, where $L$ is the length (or width, or any characteristic dimension) of the helicopter, so

$$N = kL^2\rho v^3 \quad (4)$$

where $k$ is the proportionality constant. We can write this equation for both the model and the real thing, so we are led immediately to the equation

$$\frac{N'}{N} = \left(\frac{L'}{L}\right)^2 \left(\frac{v'}{v}\right)^3 \qquad (5)$$

where $N'$ is the power required for the real helicopter and $N$ the power used for the model, $L'$ the length of the real helicopter, and so on. Equation 5 can be combined with equation 2 to give us

$$\frac{N'}{N} = \left(\frac{L'}{L}\right)^{\frac{7}{2}}.$$

Since we know that $N = 50$ watts and $L'/L = 10$, the required power is slightly less than 160 kW. Such "quick and dirty" estimates, based on scaling and dimensionality considerations, usually work very well. In this case, it required that we look at three fluxes: mass, momentum, and energy.

The roots of the helicopter are mainly Russian. In 1909, Igor Sikorsky, then 20 years old, built his first aircraft in Kiev. It was a helicopter and it did not fly well. After 30 years of great success at every other kind of aircraft, including the ancestors of the airliner, amphibian, and strategic bomber, Sikorsky, by then a U.S. citizen, tried again. He constructed the first successful helicopter in the United States in 1939.

**10. Body Heat:** Marina is quite correct. This problem was suggested by an article by Robert Emden, "Why Do We Have Winter Heating?" (*Nature* 141 [May 1938]) and a subsequent discussion by Arnold Sommerfeld (*Lectures on Theoretical Physics*, vol. 5 [New York: Academic Press, 1956]). What is in Boris's apartment? According to Boris, very little besides him; that is, there is a lot of air there. For air (or any gas at room temperature), we can write the energy per unit mass as

$$C_v T + E_o,$$

and if the average density in the apartment is ρ, the energy per unit volume is

$$U_v = \rho C_v T + \rho E_o.$$

In addition, we know the ideal gas law, $\rho = p/RT$ (where $p$ is the pressure), so the energy per unit volume is

$$U = p\left(\frac{C_v}{R}\right) + \left(\frac{p}{RT}\right)E_o.$$

If the second term is negligible (it comes from our constant when the energy was defined), then the energy of the apartment depends only on the barometric pressure.

If the second term is not (in fact, it is positive), then the energy of the apartment actually *decreases* when it is heated! Either way, Marina is correct. Where does the energy go? Literally, out the window. The pressure in the apartment remains at one atmosphere. The apartment is not sealed against gas flow, and any heating will result in an outflow of material (heated gases) and a net *decrease* in energy. On the other hand, should Marina bring in a cold bottle of vodka and leave it on the table to warm up, the increased energy of the vodka does not come from the contents of the room but rather from the outside!

In answering this question, we have neglected the flow of energy in and out of Boris and whatever contents there are in the apartment other than air. This is not a bad approximation. Boris is essentially an isothermal system with an independent energy source, and there is a *lot* of air in the apartment. As Sommerfeld points out, Boris is really paying to control the entropy, not the energy, in his apartment. The energy bills are a consequence of the control of entropy. (This concept was not realized by one of us [R.R.] until his third course in thermodynamics.) Walther Hermann Nernst, the discoverer of the Third Law of Thermodynamics, kept only fish as pets, saying (as nearly as we can translate), "to keep a warm-blooded pet is to heat the universe with your own money."

# Solutions to Chapter 7

## EXPANDING AND CONTRACTING UNIVERSES

**1. Holding the Line on the Ruble:** The line should remain straight (if the authorities have been truthful!). The equation of the line can be written, in Cartesian coordinates, as

$$Ay + Bx + C = 0.$$

From the definition of the thermal expansion coefficient, for any temperature change $\Delta T$ the coordinates should also change:

$$x \rightarrow (1+\alpha\Delta T)x$$

and

$$y \rightarrow (1+\alpha\Delta T)y,$$

so that the linear equation is transformed to

$$Ay + Bx + \frac{C}{1+\alpha\Delta T} = 0$$

This is still the equation of a straight line, with the same slope. Only the intercept is changed. Therefore the line remains straight if the heating is uniform and the material is homogeneous. There is an additional condition we might consider, and that is whether the material is isotropic, that is, whether the expansion depends on direction. Most but not all materials are reasonably isotropic as they exist in nature. However, some materials do not expand equally in all directions. For instance, suppose $\alpha$ is different for the $x$ and $y$ directions. In that case, the line will remain straight but the slope will change (you should prove this to yourself). For more complex cases, wait until you have solved the next problem. You might also consider how likely it is that the authorities have told Boris the truth. A curved line might not be out of the question after all!

**2. Inflation of the Currency:** The hole gets larger. Suppose the hole had not been drilled, but the circumference of the hole marked on the coin. From the definition of thermal expansion in the previous question, it is clear that the circumference is expanding and the radius of the circle should increase as the coin is heated, that is,

$$r \rightarrow (1+\alpha\Delta T)r.$$

Thus the hole should expand just as the material enclosed by the circle should expand. Why did Marina doubt Boris? Anyone who has baked bagels (or doughnuts, particularly ones with "holes") has seen visible evidence of this effect, where the expansion is due to the dough rising rather than to thermal expansion.

**3. Expanding Bureaucracy:** The sphere that rests on the table will require more energy. Two things will happen when the spheres heat up: the temperature will increase, and the spheres will expand. This expansion has a different effect in the two cases. For the hanging sphere, the center of gravity will be lowered because of the increase in radius of the sphere; for the resting sphere, the center of gravity will be raised by the same amount. In essence, the hanging sphere expands downward and the resting sphere expands upward, as depicted in the figure. The gravitational potential energy of the hanging sphere will be decreased, and for the resting sphere it will be increased. These energy changes must be added to the energy needed to heat the spheres up in the absence of gravity. Therefore, more energy will be required for the resting sphere.

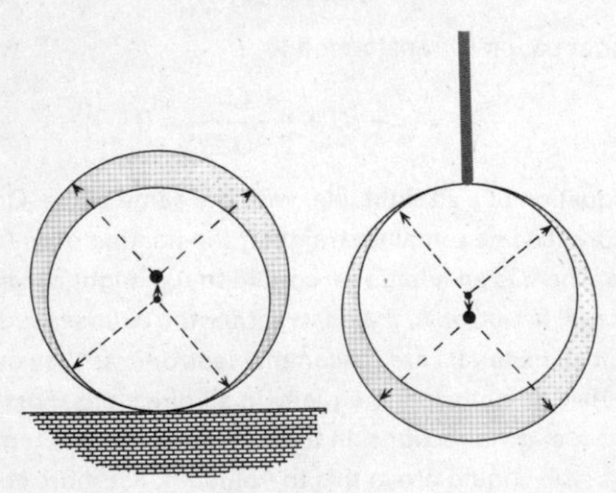

We can calculate the energy difference. As mentioned in the solutions to problems #1 and #6 in this chapter, the coefficient of linear thermal expansion $\alpha$ is defined by the relation

$$l = l_o(1 + \alpha\Delta T),$$

where $T$ is the temperature and $l$ is the length of an object or the distance between any two points on the object. We can therefore expect the radius $R$ of the spheres to increase, when the sphere's temperature increases by $\Delta T$, by an amount $\alpha R \Delta T$. The gravitational potential energy change would then be

$$mg\Delta h = mg\alpha R\Delta T.$$

The difference in energy between the two spheres would be twice this amount. Notice that the difference increases as the mass, as expected, but also as the radius of the sphere. Evidently, the effect would be larger for hollow spheres, which would have a larger radius for the same mass. Therefore, a hollow table lamp would be, by these lights, a large energy waster. It is possible that this is the reason Soviet authorities favor hanging lights from the ceiling (see "Lights Out at Malakhovka"). However, it was pointed out by Academician Zhakharovetz that this conclusion is totally false for some materials, for instance, irradiated plastics, which shrink upon heating. The ensuing furious legislative activity was a precursor to the chaos yet to come.

We will leave it to you to calculate just how large or small these bureaucratically significant effects are for a solid or hollow sphere the size of your own table lamp or any other appliance. For copper, the thermal expansion coefficient is 17.6 X $10^{-6}$ per Kelvin.

**4. Old Man Mazay Welcomes You to the Friedman-DeSitter Universe! (Is the Hubble Constant Really Constant?)**: Begin by nominating one of the hares as "most intelligent." This hare #1 will be our reference frame. The universe and everything in it move with velocity $-\mathbf{v}_1$ relative to this hare. For our Galilean transformation, we now add the velocity $-\mathbf{v}_1$ to all the velocity vectors in the original reference frame, as shown in the figure here for the third, fourth, and fifth hares.

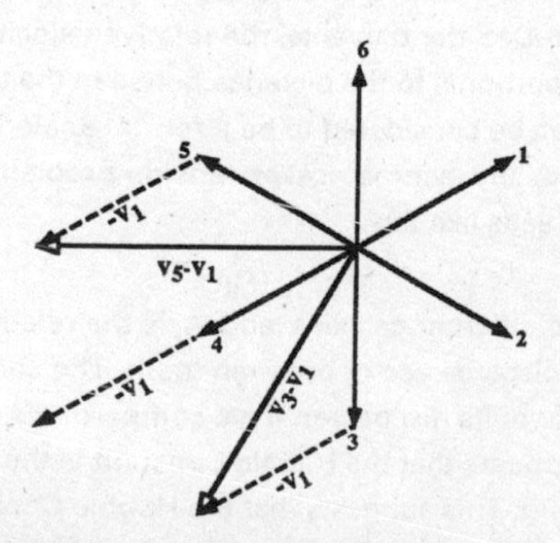

In the new frame, we obtain zero velocity (as we should) for hare #1 and new vectors for the other five, as shown in the second figure, where hare #1 sits at the center of his reference frame and watches hares #2–6 run away from him!

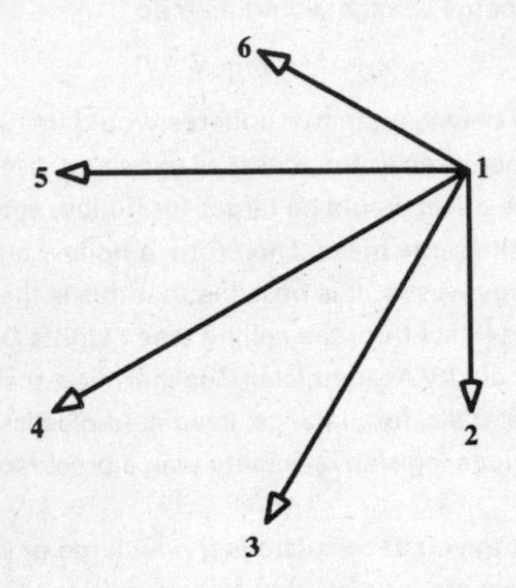


The position vectors for all the other hares are obtained from the velocity vectors by the usual equation, distance = time × velocity, so the positions in the new reference frame resemble the second figure, with each vector multiplied by the time. For hare #1 or, in fact, for any hare, $(\mathbf{r}_i - \mathbf{r}_j) = T(\mathbf{v}_i - \mathbf{v}_j)$, where $\mathbf{r}_i$ and $\mathbf{r}_j$ are position vectors and $T$ is time. The great surprise of hare #1 was due to the fact that, although he knows that all six hares are running at the same speed, it appears from his point of view that all the other hares have different velocities and different positions, and that the velocities are exactly proportional to their positions. This situation is exactly what is found for stellar objects, e.g., galaxies, and is referred to as the Friedman-DeSitter expanding universe. In the Friedman-DeSitter universe, the relative velocity vectors of any two points (galaxies) are proportional to the distance between the two points. Thus, Mazay's hare problem can be considered to be a formal analogy with the expansion of the universe or, indeed, any symmetrical expansion problem. The cosmological version of this problem reads like this:

$$\mathbf{v}_{ij} = H\mathbf{r}_{ij}$$

where i and j refer to two different galaxies, and $\mathbf{v}_{ij}$ is the relative velocity of the two galaxies and $\mathbf{r}_{ij}$ the distance vector between them. The constant $H$ is called the Hubble Constant in honor of its discoverer. If we compare this equation with the observation of hare #1, it appears that the Hubble Constant is the reciprocal of the time elapsed since the Big Bang. This suggests that the Hubble Constant is not really constant, but should decrease with time. This may or may not be true, but because human existence has occurred over such short time spans compared to the lifetime of the universe, we don't expect to see any great change.

**5. Frames of Reference and Interstellar Motion (Splitting Hares):** All the other stars in the cluster appear to be "running away" from this star as well, and again, the velocities (as observed from the second star, so they are relative velocities) are proportional to the distance from the star! Look again at the previous question and its solution. We could have solved this problem using any one of the six hares for our reference frame, and the answer would be the same! Even if the stars do not have the same speeds, if we set up a reference frame on our astronauts' star (or any star in the cluster!) and determine the relative velocities by vector addition, as we did for Old Man Mazay's hares, the astronauts will observe that the stars around them are all moving away at velocities proportional to the distances between them and the stars. We urge you to try this graphically, with different speeds. In other words, any point we choose to observe the cluster from will appear to be at the "center of the universe." We are neglecting the effects of the orbital motion of the spaceship about the "central" star, because orbital velocities and distances are tiny compared to interstellar velocities and distances.

**6. Thermal Expansion and Cosmological Cats:** As stated in this and previous problems, the linear thermal expansion coefficient $\alpha$ is defined by the equation $l = l_o(1 + \alpha\Delta T)$ between any two points. Take *any* two points $i$ and $j$ on the roof, and think of the length of the vector $\mathbf{r}_{ij}$ that connects the two points. Raise the temperature by a small amount $dT$. According to the definition of $\alpha$, the distance $|\mathbf{r}_{ij}|$ between the points must increase by the amount $dr = \alpha|\mathbf{r}_{ij}|dT$. If the rate of temperature increase is $dT/dt$ (in degrees per second), then the relative velocity, or the rate by which the distance $|\mathbf{r}_{ij}|$ increases, can be obtained from

$$\mathrm{v} = \frac{d|\mathbf{r}_{ij}|}{dt} = \frac{d|\mathbf{r}_{ij}|}{dT}\left(\frac{dT}{dt}\right) = \alpha\,\mathbf{r}_{ij}\left(\frac{dT}{dt}\right).$$

Thus the velocity between any two points on the metal sheet will be proportional to the distance between the points. If the rate of heating $dT/dt$ is constant, then the entire proportionality coefficient will be constant with time, and we arrive at a situation similar to the Friedman-DeSitter universe.

In fact, the metal roof is another example of a Friedman-DeSitter universe. The result of the uniform expansion of the roof is that all objects recede from each other at relative velocities that are proportional to the distance between them. The expansion coefficient and the heating rate have cosmological significance as well, but we will leave that exploration to you. The cats probably don't care.

**7. Masha's Mathematical Turtles:** If you follow the hint, you will arrive at the situation shown in the first figure here.

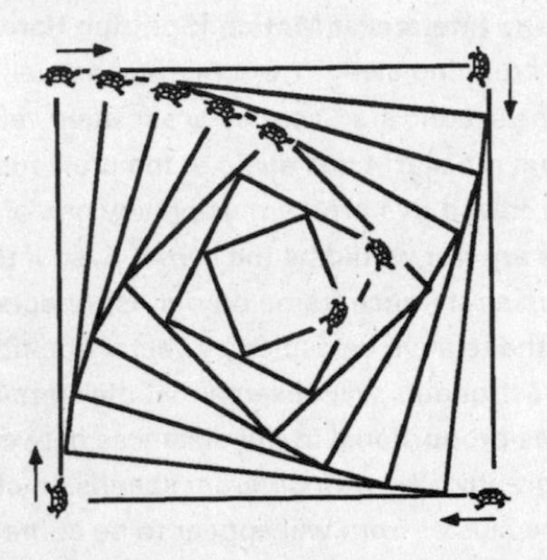

The square will rotate and shrink, and the turtles will spiral into the center of the original square. As you can see directly from the figure, the rate of the shrinkage of the side will be the turtles' constant speed v, so we can write $s = s_o - vt$ for the length of the side $s$ as a function of time, with the initial side length $s_o$. The shape does not change, but the side of the square shrinks in a linear fashion with time, and they meet in the center at $t = s_o/v$. If you wish a more formal solution with more quantitative details, look at the second figure.

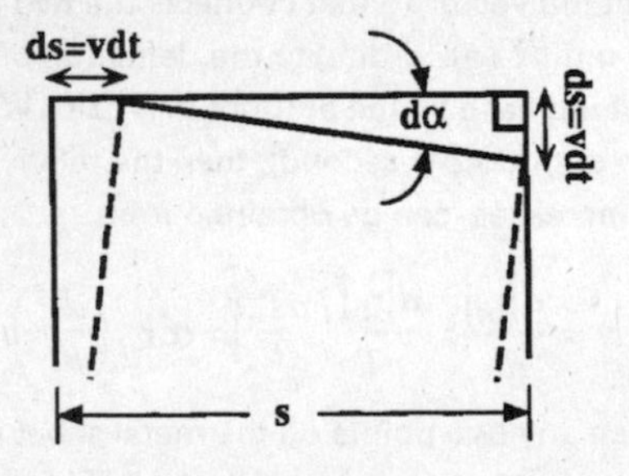


If $s$ is the side of the square, $ds \cong -vdt$ because the angle $d\alpha$ is so small (remember that the limit of $\sin\alpha$ for small $\alpha$ is simply $\alpha$). Alternatively, you can use the Pythagorean Theorem to arrive at this relationship. In either case, the result can be integrated to give $s$ as a function of time with the familiar result: $s = s_o - vt$. Now the rotation of the square can be described by the angle $\alpha$ shown in the figure. Since

$$d\alpha = \frac{-vdt}{s}$$

and the dependence of $s$ on time is now known,

$$\alpha = -\int_0^t \left( \frac{v}{s_o - vt'} \right) dt' \quad \text{or} \quad \alpha = +\ln\left(1 - \frac{v}{s_o}t\right).$$

The angle $\alpha$ "blows up" as $t$ approaches $s_o/v$, becoming infinite at that value; as slowly as the turtles may walk, the angle changes *very* rapidly as the side $s$ approaches zero! We will leave the determination of the turtles' trajectories to you. You will find that the resulting description is that of a spiral.

# Solutions to Chapter 8

## LEARNING THE ROPES

**1. Aleksandr and the Commandant:** If the water resistance is negligible and only inertia matters, Newton's Second Law can be applied. The fish scale is used to pull the boat through the water, using a constant force. Beginning with $F = ma$, we can integrate the Second Law twice to get $x = (F/2m)t^2$ for the distance moved. Aleksandr measures the distance moved with a tape measure, the time with his stopwatch, and the force applied with his fish scale. The mass of the boat (in kilograms) is then given by $m = (F/2x)t^2$. This should be done on a calm day, with no wind or waves present, and with a good stopwatch. For instance, a boat weighing 200,000 kilograms (about 200 tons) will move 1 meter in about 20 seconds using Aleksandr's fish scale. (Remember, 1 kg-force is about 10 Newtons! 100 kg force will need several sailors pulling on the fish scale!)

However, as Aleksandr's engineer knew, the mass of the boat is not the only inertial force involved. Water must be pushed around when the boat is moved. This water is also accelerated. Try to move, or even throw, any object underwater, or to walk underwater. The resistance you feel is real, and is well known to hydrodynamicists. This "virtual mass" must be added to the mass of the boat, and can amount to (depending on hull shape) as much as ⅓ of the boat mass. We can say with confidence that Aleksandr got credit for a larger catch than he really had. This kind of ingenuity in the face of failed equipment and resolute bureaucracy was common in the Soviet Union, but it has been called to our attention that it has also been used in the United States. (See the letter by Bob Gillespie and Tom McCullough on page 34 of the November/December 1994 issue, and the comment by William K. George and Roger E. A. Arndt on page 41 of the January/February 1995 issue, of *Ocean Navigator* magazine.)

**2. Unloading Aleksandr's Fish:** Kretinov is wrong again. This is apparent if his answer is examined for situations where the angle $\alpha$ is close to zero, that is, where the ropes hang straight down. Now the ropes are pulling the fish crate straight up and the velocities $u$ and v should be almost identical. However, Kretinov's solution indicates $u$=2v, which is obviously incorrect. Kretinov has confused this problem with calculation of the *forces* on the crate and the ropes, which *is* done by adding vector components. However, a careful examination of the figure shown with the question should, due to the extremely liberal use of the "vector" **v**, make you very suspicious of this approach for this problem. In fact, we must calculate speeds or at least displacements, not forces, to answer this question. Vector resolution and addition is not the answer here. Because each rope and pulley must behave (we are told) in an identical manner, the problem can be solved for one side of the system, as shown in the figure.

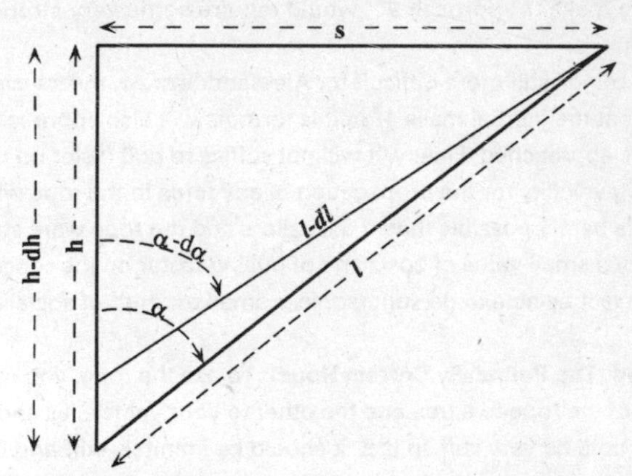


The horizontal distance $s$ between the pulley and the fish crate remains constant, but the length of rope $l$ between the pulley and the fish crate shortens at the constant rate v. This means that the vertical distance $h$ and the angle $\alpha$ must change. We have a right triangle with two sides ($h$ and $l$) and one angle ($\alpha$) changing. We can write for this right triangle the Pythagorean relation

$$s^2 + h^2 = l^2$$

and differentiate it:

$$2hdh = 2ldl$$

or

$$\frac{dh}{dl} = \frac{l}{h} = \frac{1}{\cos\alpha}$$

To get the velocity with which the crate rises, we write

$$u = \frac{dh}{dt} = \left(\frac{1}{\cos\alpha}\right)\frac{dl}{dt} = \frac{\mathrm{v}}{\cos\alpha}.$$

**3. Aleksandr's Fish Go Supersonic:** No, it is not possible. If it were possible to actually make $\alpha$ approach 90° *and* the ropes were made of a *very* special material, then we would be more optimistic. To begin with, as $\alpha$ approaches 90°, the forces on the ropes would approach infinity. Kretinov's approach can be used for the calculation of the *forces* on the ropes, which would be given by the formula $Mg=2T\cos\alpha$, where $M$ is the mass of the fish crate, $g$ the gravitational constant, and $T$ the tension in the ropes. Clearly, if this equation is solved for $T$, we find that $T$ approaches infinity as $\alpha$ approaches 90°. To make $\alpha$ approach 90° would require some very strong rope, and very strong sailors!

To make things still more difficult for Aleksandr's crew, the special theory of relativity predicts that the inertial mass $M$ in this formula will also approach infinity as the speed of light is approached. Finally, it will not suffice to pull faster on the ropes, because the limiting velocity for the propagation of any force in the rope will be the speed of sound. It is barely possible that, if the sailors and the rope were strong enough, by a combination of a small value of $\cos\alpha$ and of pulling faster on the ropes, Aleksandr's fish may in fact be able to go supersonic, a small triumph of socialism.

**4. Stuck in the Mud (The Politically Correct Rope):** To use the rope, you must begin by fastening one end of the rope to a tree and the other to your vehicle, as shown in the figure. The rope should be very stiff. In fact, it should be infinitely stiff and infinitely strong if infinite forces are to be applied. Because the rope has been specified as such in an official directive, we will assume that it will have these remarkable characteristics.

As the figure suggests, you should *push* on the rope, at or near the center of the span. The rope will pull the truck toward you, as shown. This is a "simple machine," and we can use the principle of virtual work to determine the forces involved. The "input work" you will do if you push on the rope is just $fS$, where $f$ is the force you have exerted and $S$ is the displacement of the center of the rope. The "output work" done by the machine is $Fs$, where $F$ is the force on the truck and $s$ is the displacement of the truck. The virtual work principle requires that at static equilibrium, the net virtual work sums up to zero, in this case $fS=Fs$. The mechanical advantage of the machine $F/f$ can now be determined: $F/f=S/s$. We must determine the ratio $S/s$. If the rope is indeed infinitely stiff and does not stretch, we can see from the bottom of the figure that the Pythagorean Theorem can be used in the following way:

$$L^2 = S^2 + (L-s)^2 = S^2 + L^2 - 2Ls + s^2 .$$

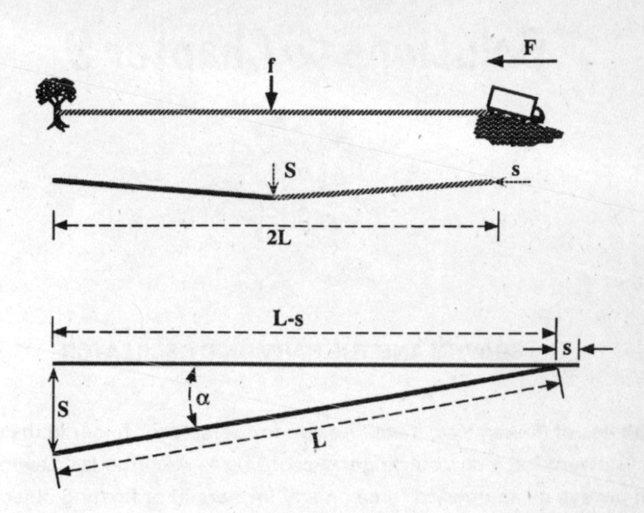


In general, the displacement $s$ of the truck must be small; we can see that the angle between the two positions of the rope will be small, and furthermore if $s$ is not small the mechanical advantage $F/f=S/s$ will not be large. Therefore, we will ignore $s^2$, the last term on the right. Now the equation can easily be solved to obtain the mechanical advantage:

$$\frac{S}{s}=\frac{2L}{S}$$

Now, if the rope is very long ($L$ will be large) and we displace the center of the rope by a small amount ($S$ will be small when compared with $L$), the mechanical advantage can be very large, and as $S$ approaches zero the mechanical advantage approaches infinity, as the authorities have claimed. (That is, with a politically correct rope, infinite force is achievable.) Of course, the displacement $s$ of the truck will be even smaller than $S$. You will need to pull the truck by this tiny amount, reblock the wheels so it does not slide back, refasten the rope, push the center again—you will be there for a very long time. You could wait for the mud to freeze, but it would probably also snow, bringing back your problem in another form. Your best alternative is to call the AAA.

Of course, all this is conceivable only because the rope provided by the Party is infinitely strong and infinitely stiff. If the rope stretches at all, the displacement $s$ of the truck will practically vanish. The triumph of theory demands such exceptional materials as a rope that will not stretch or fail, even with infinite force.

# Solutions to Chapter 9

## GRAVITY AND THE HARMONIC OSCILLATOR

**1. The Balance of Power:** Yes, it will. As you immerse your finger in the water, there will be an upward force on your finger. According to Archimedes' Law of Buoyancy, there will always be an upward force on any immersed or floating object, equal to the weight of the water displaced by that object. By Newton's Third Law, your finger must exert a force that is equal in magnitude but opposite in direction on the water. This force is then transferred to the glass and then to the scale.

Archimedes' Law of Buoyancy is a remarkable example of a physical law that is valid to an extremely high degree of precision and can be easily derived theoretically from basic concepts of mechanics. The brilliant seventeenth-century proof of this law by Blaise Pascal is one of the most elegant chains of reason in the history of the sciences. Here is that proof:

Remember that if a body is in equilibrium (not accelerating), the sum of the forces acting on it must be zero. No other basis is needed. Consider a fully immersed body, as shown in the figure.

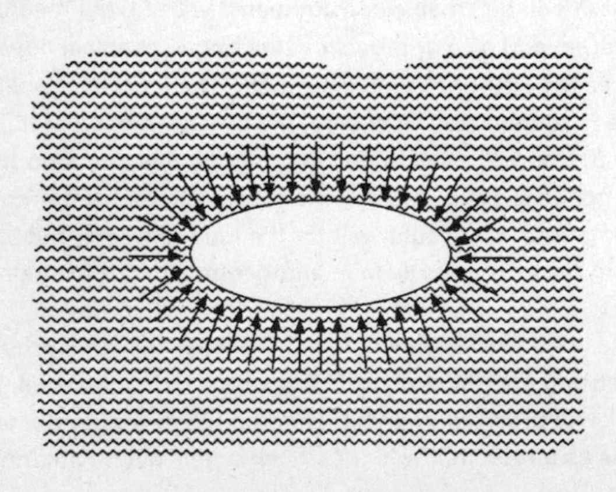

Clearly, the surrounding water can act on this body only at its surface. The Archimedes force is the sum of all the forces acting on all surface elements of the body, and these elementary forces cannot know what is inside the body. In fact, if we could remove the entire body and replace the volume occupied by the body with water (or anything else!), the Archimedes force should not change. However, we know that if the body were to be replaced by a mass of water of the exact same shape and size, the body would definitely be in mechanical equilibrium with the water around it. Therefore, the weight of the body of water is exactly compensated by the forces from the water around it, i.e., the Archimedes force. Consequently, the Archimedes force, the total force on any immersed body, must be equal to the weight of an equivalent volume (referred to as the "displaced volume") of water. The force depends only on the displaced volume, not on the shape or constitution of the body. This proof can easily be extended to floating objects, by the same reasoning.

The Archimedes force was part of everyday life in Russia, where prepackaged foods and bar codes do not yet exist. Goods are usually weighed on scales in the presence of the customer, and this situation presents many opportunities for cheating by the salesperson. An integral part of the Russian art of selling sour cream is the surreptitious finger in the product during weighing. The total gain, perhaps half an ounce, was taken as proof of the superiority of the salesperson. In public eating places, sour cream is professionally diluted, not by fingers or milk or yogurt, but by paper napkins. Stirred into the sour cream in the evening, they dissolved by morning and left the sour cream thicker and heavier. When Y.C. discovered this procedure, it immediately resolved another question that had bothered him for years: why paper napkins were never available in cafeterias.

**2. Vanya and the Water Bucket (A Matter of Location):** No. Remember Archimedes' Law: the total weight of the ice floating in the bucket must be equal to the weight of the displaced water. Melting will certainly not change the mass (or weight) of the frozen water, and so the volume of the molten ice will exactly match the volume of water displaced by the ice! Therefore, melting will not by itself cause overflowing of the bucket. However, if we are to keep Vanya's mother happy, there are other concerns, because melting is not the only cause of volume change. Water is remarkable in that it has maximum density at 4° C. Between 0° C and 4° C, water actually shrinks as the temperature increases. At temperatures above 4° C, water expands, like almost everything else. Therefore, on melting, the bucket will not overflow, and as the water heats up the water level will first go down, and then up. If the entrance hall is warm enough, thermal expansion will cause overflow. In Russia, where the entrance hall (called *seni*) is unheated and not much warmer than 0° C at all times during the winter, thermal expansion of anything at all is not a problem.

**3. Towing the Barge:** No, you should not! The gravitational work, and the gravitational force, can be ignored here. The gravitational force is always compensated by the buoyancy (Archimedes) force, for any floating object. A surprising (and counterintuitive) consequence is that the same power is required to tow the barge upstream as downstream, provided that the velocity through the water is the same! (In practice, this is never the case!)

**4. Beam Me Up, Scotty!:** No, it will not work. In fact, the astronauts will be in even greater difficulty. The accelerations due to gravity or due to the motion of the spaceship act on every element of mass in the astronaut's body and result in the force $m(a + g)$, regardless of immersion. The buoyancy forces act on the surfaces of the body only, and will not prevent the effects of acceleration. If anything, the astronauts will be in even greater discomfort due to the lack of proper supporting structures. In addition, the hydrostatic forces due to the increase in effective gravity will increase. If the astronaut is under 2 feet of water and the overall acceleration is 20 g's, the pressure will be the same as 40 feet and 1 g.

**5. Fixing the Clock (A Grave Matter):** The period $T$ of the pendulum is given by the equation

$$T = 2\pi\sqrt{\frac{l}{g}}$$

where $l$ is the length of the pendulum and $g$ is the gravitational acceleration. The gravitational acceleration varies with position on the Earth's surface due to density variations of the Earth and, even more important, variations in centrifugal acceleration. The constant $g$ is really the sum of two effects: gravitational and centrifugal. The centrifugal component is of the order of 1/1,000 of the gravitational component. The proper component of the centrifugal acceleration must be subtracted from the true gravitational acceleration. Because locations near the equator are farther from the axis of rotation of the Earth, the centrifugal acceleration is greater, and therefore the apparent gravitational acceleration $g$ is less than it would be in France. Therefore, the period $T$ of the pendulum would be too long in North Africa if it were properly adjusted for France. The difference in $T$ is quite noticeable. If $T$ is 1 second, the clock will lose about 1 minute per day (depending on exact latitudes). The effective length $l$ of the pendulum must be shortened in order to make up for the decrease in $g$, and this is what Huygens advised his friend to do. Huygens invented several features to improve accuracy, including the escapement and the cycloidal pendulum, but he had not anticipated, at least at that time, variations in $g$. The same effect will occur due to local mass concentrations. For instance, a small iron ore deposit (hematite, magnetite, or taconite, it doesn't matter, about 10 cubic kilometers) will have the same effect.

**6. The Clock in the Elevator:** The period $T$ of the pendulum is inversely proportional to the square root of the gravitational acceleration, that is, $T \propto g^{-1/2}$. The effective gravitational acceleration increases when the elevator accelerates upward and decreases when the elevator accelerates downward. It is clear, then, that the clock will go faster when the elevator accelerates upward (the period of the pendulum will decrease) and slower when the elevator accelerates down. We need to analyze the gains and losses in order to determine the net change. For acceleration upward we can write $T_u \propto (g+a)^{-1/2}$, and for acceleration downward we can write $T_d \propto (g-a)^{-1/2}$. For each trip of the elevator, regardless of direction, there will be one upward acceleration and one downward acceleration of equal duration amounting to a total time $2t_o$, because the trip begins and ends with the elevator at rest. The time $t'$ counted by the clock in the elevator during the upward acceleration will be $t'=t_o(T_o/T_u)$, where $T_o$ is the period of the clock when it is at rest or in uniform motion, and $T_u$ is the period of the clock when it is under upward acceleration. Likewise, when the acceleration is downward, we can write $t''=t_o(T_o/T_d)$. We can now compare the total time recorded by the clock on the elevator:

$$t' + t'' = t_o[(1+a/g)^{1/2} + (1-a/g)^{1/2}]$$

with the time $2t_o$ measured by the office clock. It turns out that the expression inside the brackets is always less than 2 when the accelerations are less in magnitude than the gravitational acceleration $g$. (We offer alternative proofs of this statement at the end of this solution.) If the bracketed value is less than 2, then $t' + t'' \leq 2t_o$, and the elevator clock will show less time than the office clock. The operator is underpaid.

Here are our proofs, where in each case we let $x = a/g$:

For real $x$, if

$$x^2 > 0 \text{ then } 1-x^2 < 1,\ 2(1-x^2)^{1/2} < 2, \text{ and } (1-x)+(1+x)+2(1-x^2)^{1/2} < 4.$$

Now the left side of the last inequality is equal to

$$[(1+x)^{1/2} + (1-x)^{1/2}]^2,$$

which you can verify by performing the squaring operation. Now take the square root of both sides and you will see that the quantity in brackets is always less than 2.

Since $x$ must be smaller than 1 (we agreed to limit the acceleration of the elevator to less than $g$ in order to avoid bouncing passengers off the ceiling), you can also use the Taylor series to prove the inequality.

**7. The Candle on the Carousel:** The candle flame will lean inward, toward the center of the carousel. In order to solve this problem as well as the previous problem, you must concern yourself with inertial forces. In the most elementary applications of Newtonian dynamics, you have probably learned to sum up the forces exerted on a body; if the sum is zero, the momentum does not change. Such inertial frames of reference are straightforward. The previous problem, with a pendulum in an elevator, is difficult to analyze within an inertial frame because of the acceleration of the elevator. Frames that accelerate relative to the Earth *usually* are noninertial frames of reference. Consider two reference frames, one translating relative to the other with acceleration $\boldsymbol{a}_o(t)$. Measured in the laboratory (inertial) frame, any acceleration $\boldsymbol{a}$ would be equal to the sum of the acceleration $\boldsymbol{a}'$ measured in the accelerating frame plus the acceleration $\boldsymbol{a}_o(t)$. That is,

$$\boldsymbol{a} = \boldsymbol{a}' + \boldsymbol{a}_o(t).$$

This relation follows from the identity $\boldsymbol{r} = \boldsymbol{r}' + \boldsymbol{r}_o$ for the two frames; simply take time derivatives. Now if Newton's Law

$$\sum F_i = ma$$

is used in the laboratory frame, we obtain

$$ma' = -ma_o + \sum F_i.$$

Now $\boldsymbol{a}'$ is the acceleration in the new, noninertial frame, and it is clear that Newton's Law does not hold any more for $\boldsymbol{a}'$. We have to add a new "artificial" force $-m\boldsymbol{a}_o$ to the sum of the forces. This force, called "the inertial force," has meaning only in the accelerating frame and must always be added in noninertial frames to the real, tangible forces.

Now our candle flame is rotating with the carousel with angular velocity $\omega$ and therefore has an acceleration $\omega^2 R$, where $R$ is the distance from the center of the carousel. The candle is resting relative to the carousel, but the rotation of the carousel and the resulting acceleration give rise to a centrifugal inertial force. The absolute value of this force is $ma_c = m\omega^2 R$. For any mass, the acceleration would be directed outward. However, the candle flame is subjected to buoyancy forces, as it is lighter than the colder air around it. Therefore, the flame leans inward, toward the center of rotation.

The most important feature of inertial forces is that they act on all elements of the body and are always proportional to the mass of the body. For this reason, Einstein declared inertial forces to be equivalent to gravitational forces. This statement, the "equivalency principle," became the cornerstone of general relativity (Einstein's Theory of Gravitation).

**8. Why Does the Moon Cause Tides?:** The Earth is large compared to the moon and is significant even compared to the distance to the moon. The attractive force is directed toward the center of the moon and is (according to Isaac Newton) inversely proportional to the square of the distance between any point on the Earth and the moon's center. However, the side of the Earth facing the moon is 4,000 miles closer to the moon than the center is, and 8,000 miles closer than the opposite side. This difference in attractive gravitational forces is itself a force that causes tides. More precisely, the tidal force is caused by the gradient in gravitational attraction and, as such, it varies approximately inversely as the *cube* of the distance to the center of the moon. (You can demonstrate this with calculus and Newton's Law of Universal Gravitation if you wish; a more vivid illustration can be found in Larry Niven's science fiction short story "Neutron Star.")

**9. Gravitational Attraction at the Breakfast Table:** Newton's Law of Gravitation states that for any two *concentrated* masses, the attractive force between them is inversely proportional to the square of distance between them:

$$F(R) = G\frac{Mm}{R^2}$$

where $G$ is the gravitational constant, R the distance between the masses $M$ and $m$, and $F(R)$ the attractive force. Thus for two pointlike objects, or at least objects of simple geometry, the attractive force should increase as they are brought together. Now we are going to consider a slightly more complex situation, the gravitational forces between a concentrated mass $m$ (say, a small sphere) and a ring, as shown in the figure.

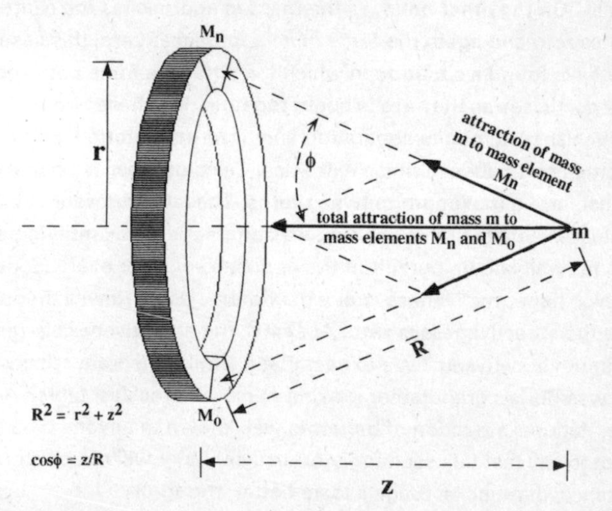


Clearly, if the sphere is positioned at the center of the ring, the gravitational force between the two masses is zero due to symmetry: the attractive force on the mass $m$ from any piece of the ring will be canceled by an equal and opposite force from an identical piece on the other side of the ring. On the other hand, when the distance between the ring and the sphere is nearly infinite, the force is also approximately zero due to the distance.

Now do the following. In your mind (looking at the figure while you do this will help), construct a line normal to the ring, running through the center of the ring and running out to infinity. Move the small mass $m$ along that line. We can analyze that situation precisely. Look at the figure: by symmetry we can see that the direction of the net force on the mass $m$ due to the ring is directed toward the center of the ring. Consider the forces due to two small pieces of the ring having equal mass $M_n = M_o$, on opposite sides, as shown in the figure. Using Newton's Law and adding the two force vectors, we find that the net force is

$$F(R) = 2G \frac{M_n M}{R^2} \cos\phi = \frac{2GM_n mz}{(r^2 + z^2)^{\frac{3}{2}}}$$

To find the total force exerted by the entire ring, we could consider $M_n$ and $M_o$ as differential masses and integrate around the entire ring. To answer grandfather's question, however, we need not do that.

Observe how the attractive force $F(R)$ changes as the sphere moves away from the ring: the angle $\phi$ approaches zero, $\cos\phi$ approaches unity, and $F(R)$ resembles Newton's Law for concentrated masses. The force also approaches zero as $R$ and $z$ approach infinity. On the other hand, as the mass $m$ approaches the center of the ring, $z$ approaches zero and again the force $F(R)$ approaches zero, this case vanishing at $z = 0$. We have found a situation in which the attractive force between two masses actually decreases as they are brought together. For these two bodies, the attractive force vanishes at infinite separation and at zero separation, and has a maximum at some point between these extremes. Calculus teaches us that any positive function must have a maximum between zeros. The late Professor Lev Landau, one of the greatest scientists of this century, was infamous for quantifying everything, including his well-known pursuit of the opposite sex: "For every woman there's a distance at which her attractiveness is at a maximum. Using Rowl's theorem, at infinite distance the attractiveness is zero. At $D = 0$, the attractiveness is zero. There must be a maximum in between." We expect that a similar theorem applies to men.

What was Boris's grandfather looking at on the breakfast table? A bagel, of course. The gravitational attraction of bagels is well known to anyone who has eaten them. This is not to say that this seemingly bizarre behavior will not occur for other shapes, for instance, dumbbells. (Bagels taste better, though.)

**10. Lights Out at Malakhovka (The Harmonic Oscillator):** The effect is real and not due to the vodka. A lamp (or any other object!) hung by a spring of any kind in the

Earth's gravitational field will undergo simple harmonic motion. Evidently, the friction is quite low in this case, as the oscillation was easy to cause and continue. For any harmonic oscillator, the kinetic and potential energy vary through the cycle. At the midpoint, the kinetic energy (and therefore the velocity) is maximum. At the ends of the cycle, the potential energy is at a maximum and the kinetic energy is zero. Thus the velocity of the lamp will reach its maximum at the midpoint of its path, and the lamp will slow down as it approaches the end points. At each end point, the lamp will change direction, briefly having no velocity and then accelerating to the midpoint. Because the velocity of the lamp is lowest at the ends of its path, the lamp spends a longer amount of time in the vicinity of the end points than it does at any other points in its motion.

Now, what do we mean by *brightness?* We are referring to the rate at which energy is emitted from a specific surface area. Divide the path of an oscillating lightbulb into equal segments; the bulb's width makes these segments into equal areas. The bulb spends the most time in the segments at the end points because its velocity is lowest there. Because the bulb emits energy at a constant rate as a function of time (e.g., 100 watts), the most energy will be emitted from the end point segments and therefore these will appear to you to be the brightest.

In order for you to notice the changing intensity of the light, you must be reasonably close to the oscillating lamp. Thus, in calculating the most opportune time in which to hit the lamp, we can ignore the time lag between an event occurring and your eye registering that event. Since the velocity of the lamp reaches zero at its end points, this would be the ideal time in which to hit the lamp. So, in order to have the highest probability of hitting the lamp, you should fire your rifle when you see the lamp begin to become brighter and aim for one of the end points.

The nineteenth-century German physicist Hermann Ludwig Ferdinand von Helmholtz, a major contributor to the theories of electromagnetism, thermodynamics, and optics, was also the father of ophthalmology. He observed the same effect using a swinging lamp, and found not only that the lamp appeared brighter at the ends of the swing but also that there was a color effect: the light appeared white only at the ends, with colored streaks in the middle of the swing. The colors appear because the velocity of the lamp is highest at the middle of the swing and the color-sensing elements in our eyes have different response times, depending on the color. Thus, if the lamp swings fast enough, the "slowest" colors will not be sensed.

**11. The Platinum Planet:** In a low circular orbit the distance from the planet's center is, within a few percent, very well approximated by the planetary radius $R$. The centripetal acceleration $\omega^2 R$ (where $\omega$ is, as usual, the angular velocity) is balanced by the gravitational force, which is given by Newton's Law of Gravitation, $GM/R^2$. ($G$ is the gravitational constant and $M$ is the planetary mass. If we equate these two forces, we obtain

$$\omega^2 R = \frac{GM}{R^2}, \text{ or } \omega^2 = \frac{GM}{R^3}$$

Now the volume of the planet is proportional to $R^3$, and the term $M/R^3$ is proportional to the density of the planet. Therefore, the square of the angular velocity is proportional to the density of the planet, with a universal proportionality constant (for planets of approximately the same shape!). If the Earth's low orbit satellite angular velocity is $\Omega$ and the Earth's density is $\rho_e$, then we have

$$\frac{\omega}{\Omega} = \sqrt{\frac{\rho}{\rho_e}}$$

The orbital period $T$ is related to the angular velocity by $\omega = 2\pi/T$, so this equation can be rewritten as

$$\rho = \rho_e \left(\frac{T_e}{T}\right)^2$$

Now, if the orbital period is half that of Earth for low orbit, the density must be 4 times as great, or 22 $g/cm^3$. As you can see, the density of the planet may be measured with nothing more than a kitchen timer! (This density is actually slightly higher than the density of platinum, but the crew members were quite familiar with this cook's methods: he overcooked everything.)

**12. All Tunnels Lead to Moscow:** Nobody wins! The tunnels would be isochronal, with *exactly* the same transit time. Consider the situation at point A in the figure. As mentioned in our hint, the shell of mass with radius greater than *r* exerts no gravi-

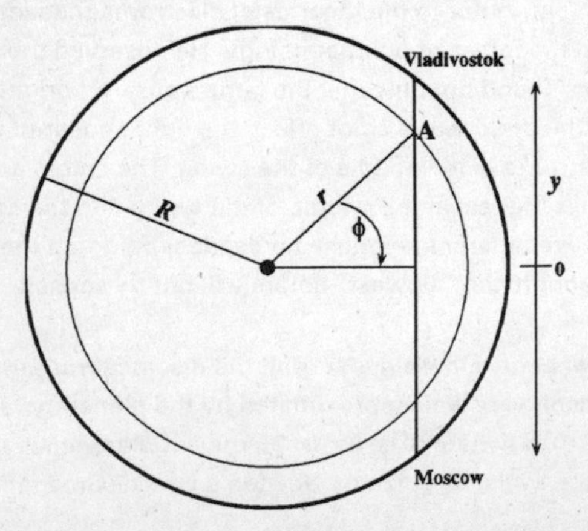

tational force whatsoever on our parcel in the tunnel. The gravitational force is then due to the spherical mass with radius $r$. The force is the same as if the whole mass were concentrated in its center. Denoting by $\rho$ the density of the Earth, we can write the mass of a sphere with radius $r$ as $M(r) = 4\pi\rho r^3/3$. According to Newton's Law of Gravitation, this mass attracts any unit mass positioned on its surface with the force $g(r) = GM(r)/r^2 = 4\pi G\rho r/3$. We have denoted the force on a unit mass by $g(r)$ because it is actually the gravitational acceleration on the surface of such a sphere. (In particular, when $r = R$, $g(R) = g$.) It allows us to write $g(r)$ in a simple form

$$g(r) = \frac{gr}{R}$$

Now we want to consider a motion of a parcel along the tunnel under the effect of such a force. We introduce the $y$-axis and angle $\phi$, as shown in the figure. The acceleration along the tunnel is $g_y = -g(r)\sin\phi$. On the other hand $y = r\sin\phi$. This, together with the equation above for $g(r)$, finally yields

$$g_y = \frac{-gy}{R}$$

*This is the equation of a linear oscillator with the "elastic constant" $g/R$.* The restoring force is proportional to the deviation from the equilibrium point $y = 0$ (the middle of the tunnel). If not picked up, the parcel will oscillate between the initial and destination points. At the initial point, at the surface the parcel possesses only potential energy. As it is accelerating along the tunnel, the potential energy is being transformed into kinetic energy, which reaches its maximum in the middle of the tunnel. When the body reaches the other end of the tunnel, the potential energy must be the same as in the initial point and the kinetic energy zero. The final velocity at either end of the tunnel is therefore always zero. If the parcel is not caught right away, it will start its way back to its senders and will thereby complete the first period of motion. The duration of a one-way trip for the parcel is half the period of the oscillator. The period $T$ of such an oscillator is independent of the amplitude, which means that the delivery time will be the same for all the Soviet cities. The oscillation cyclic frequency $\omega = 2\pi/T$ is readily expressed through the "elastic constant" $g/R$ as

$$\omega = \sqrt{\frac{g}{R}}.$$ It allows us to write the period as $$T = 2\pi\sqrt{\frac{R}{g}}.$$

If you recently solved the previous problem, you may be interested to know that this period is equal to the period of a satellite in low orbit about the Earth. You can easily calculate ($R = 6{,}400\text{km}$, $g = 10\text{m/s}^2$) that $T = 84$ minutes and the delivery time will be 42 minutes.

# Solutions to Chapter 10

## MECHANICS AND RELATIVITY

**1. Crossing Swords:** Choose the frame of reference of an outside observer. In this reference frame, the two bars are moving symmetrically, as shown in the figure.

In the figure, $\mathbf{u}$ represents the velocity of the point of intersection, $O$ represents the initial intersection point, and $O'$ indicates the point of intersection after some time $\Delta t$. Because each bar is moving perpendicular to itself with the same speed, the symmetry of motion produces a rhombus, where the diagonal $OO'$ is a bisector. Due to the symmetry, $OO'$ represents the trajectory of the intersecting point. The height of the rhombus is $\Delta h = \mathrm{v}\Delta t$ and the angle at the vertices $O$ and $O'$ is $2\alpha$. The height of the rhombus is then given by

$$\Delta h = \Delta s \sin \alpha$$

where $\Delta s = |OO'|$ is related to the velocity $\mathbf{u}$ by the equation $\Delta s = \mathbf{u}\Delta t$. Combining these equations, we obtain

$$\mathbf{u} = \mathrm{v} / \sin\alpha$$

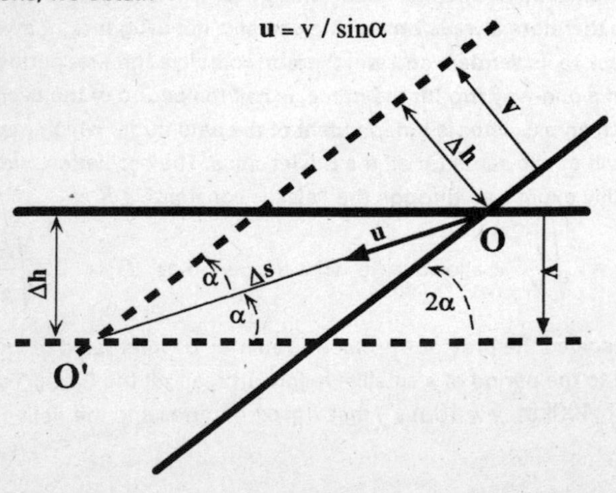

The velocity is so high because, in the swordplay, the angle $\alpha$ is usually very small. We have also simplified the problem by assuming only perpendicular translation. You will find, after reconsidering this problem, that longitudinal motion (i.e., stabbing rather than slashing) does not change the answer.

**2. Special Relativity in the Tailor Shop (Sasha Exceeds the Speed of Light):** Masha is correct again. Special Relativity has not been violated. Material objects may not move at speeds faster than that of light. The intersection of the scissor blades is not a material object. However, there is a second question of interest here. Can Sasha cut materials at super-light speeds? To answer this question, look at the first equation for $s$, and you will see that $s$ is a meaningful physical quantity only if it is less than the length of the scissor blades. Sasha will be able to make major advances in the garment industry only if he has a *very* long pair of scissors! On the other hand, it is clear that very high cutting speeds can be practically achieved.

**3. Bowling Alleys:** The lower ball will reach its destination first. Suppose, to begin with, that the balls slide rather than roll. Conservation of energy requires for each ball that

$$\frac{1}{2}mv^2 + mgh = \frac{1}{2}mv_o^2$$

where $m$ is the mass of a ball, v is the velocity, $h$ the altitude (measured, say, from the flat portion at the beginning of the run), and $v_o$ the initial velocity. Clearly the lower ball will have equal or greater kinetic energy at all points, compared to the upper one, because its potential energy $mgh$ will be equal or less than that of the upper ball. If the balls roll and do not slide, there will be an additional kinetic energy term, the rotational kinetic energy. However, the rotational velocity $\omega$ and the translational velocity are related: $v = \omega r$, where $r$ is the ball radius. Consequently, a greater total kinetic energy at all points still means a greater velocity, and the lower ball still gets there first. Even if friction is present and some combination of rolling and sliding occurs, the lower ball still gets there first.

**4. Boris and the Locomotive:** The analysis is correct. It simply does not go far enough. The force from the push rod would indeed make the locomotive move backward, but that is not the only force on the wheel. There is another moment acting on the wheel, about point $A$. The moment comes from the force exerted on the wheel by the axle, which pushes the wheel forward. This force causes a greater moment about point $A$, in the clockwise direction, and the locomotive moves forward. Where does

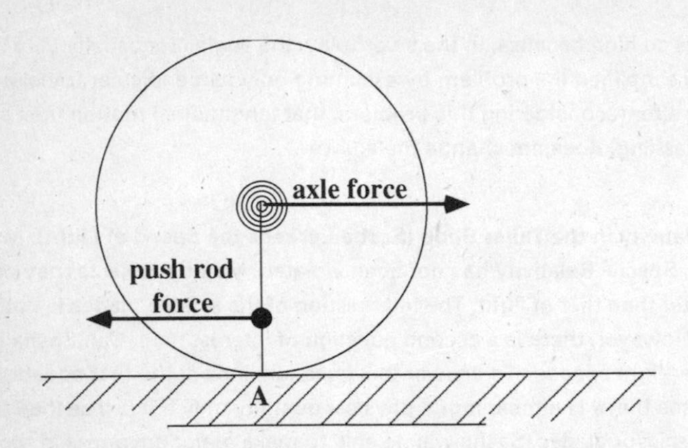


this force come from? The steam pressures on the piston and the front wall of the cylinder are equal, and the forces are also equal in magnitude, but in opposite directions. Because the cylinder is rigidly attached to the frame of the locomotive frame, this force is transferred to the axle(s) and to the centers of the wheels. Because this latter force has a greater lever arm about point $A$ than the push rod force, the total moment is clockwise. Boris was correct this time, but did not carry the analysis far enough.

**5. The Ball Game:** Put both spheres on a ramp or incline. When released, the copper sphere will roll down more slowly, and the difference will be large enough to notice easily. Surprised? Yes, if you dropped them off a tower à la Galileo, they would fall at exactly the same rate, but they will roll down an incline at *different* rates. Read on!

The difference in behavior is due to different *moments of inertia* for the two spheres. You may recall that when Newton's Second Law ($F = ma$) is applied to rotation, the mass moment of inertia measures the inertial resistance to rotation in a manner completely analogous to the role played by the mass when translation is involved. Now, the total torque $\tau$ is given by the product of the mass moment of inertia $I$ and the angular acceleration $\alpha$, that is, $\tau = I\alpha$. The total kinetic energy of a translating, rotating body is $\frac{1}{2}mv^2 + \frac{1}{2}I\omega^2$, where $\omega$ is the rotational velocity. The distribution of mass is quite important to the moment of inertia, and the distribution of mass in the two spheres is quite different.

Copper is more than three times as dense as aluminum, and therefore the shell of copper surrounding the hollow interior must be much thinner, if the spheres are to have the same mass. The figure would be typical of our two spheres, for copper and aluminum.

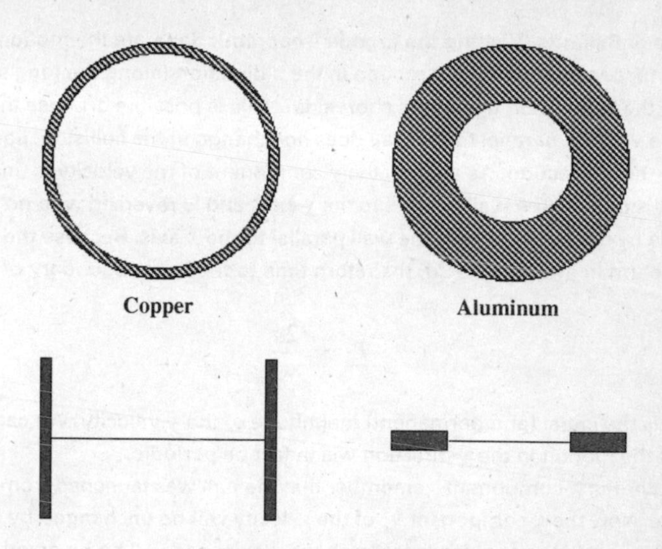


Because the copper is, on the average, farther away from the axis of rotation, the *moment of inertia* of the copper sphere is larger. If the spheres in the illustration were 2 cm in diameter, for instance, the difference would be about 40%. Now consider the difference in behavior when the spheres roll down the plane. At any point on the plane, a certain quantity of potential (gravitational) energy will have been converted to kinetic energy, and both spheres must have the same kinetic energy when they pass that point. However, the kinetic energy is now the sum of two terms, translational and rotational. Because the moment of inertia $I$ is larger for the copper sphere, the rotational velocity (and consequently the translational velocity, since the sphere is rolling down the incline) must be less. Consequently, the aluminum sphere will reach the bottom first. For our 2-cm spheres, the velocity difference should be about 6%, easily noticeable.

The distribution of mass is what determines the moment of inertia. The contribution of any mass element is equal to the product of the mass and the square of its distance from the center of rotation. To develop an intuitive feel for this, look at the two "dumbbell" analogs sketched at the bottom of the figure, for the two spheres. Most of the mass of the copper dumbbell is concentrated at the very ends of the bar, whereas the aluminum mass is distributed over much of the length of the bar. Which one will be easier to twirl? If you are not in the mood (or, more likely, lack the skill) to twirl a baton, consider shoveling—in fact, consider long- and short-handled shovels. Which one will require more effort to manipulate, and why? If you are tired of shoveling, try a sledgehammer, with a short or a long handle. Which requires more torque to swing?

**6. A Game of Billiards (Visiting the Ergodic Theorem):** Separate the motion of the ball into two *independent* problems: motion in the $x$-direction (along the long side) and motion in the $y$-direction (along the short side). This is possible because the component of the velocity parallel to the wall does not change in the collision. Look at the motion in the $y$-direction. As agreed, the $y$-component of the velocity is unchanged by the collision with the wall parallel to the $y$-axis, and is reversed with no loss in magnitude by the collision with the wall parallel to the $x$-axis. Because the total distance to return in one cycle is $2a$, the return time to the lower boundary of the table is then

$$T_a = \frac{2a}{v_y}$$

where $v_y$ is the initial (and permanent) magnitude of the $y$-velocity. We can see at this point that the motion in the $y$-direction will in fact be periodic.

For the $x$-component, remember that the ball was launched from the middle of the side. Now the $x$-component $v_x$ of the velocity will be unchanged by collision with the upper and lower sides (parallel to the $x$-axis) and will be reversed by collision with the left and right sides. The time of arrival back to the original $x$-coordinate is then given by

$$T_b = \frac{2a}{v_x}$$

In order for the ball to return to its original starting point, its $x$ and $y$ positions must simultaneously coincide with the starting point, which will occur after some number of collisions with the sides if

$$mT_a = nT_b$$

where $m$ and $n$ are integers. These three equations can be combined now:

$$\frac{v_y}{v_x} = \tan\phi = \frac{m}{n}$$

where $\phi$ is the angle shown in the figure with the original question. Any launching angle that satisfies this equation will gurarantee that the ball will eventually arrive again at the point where it was launched. This is an amazing result: *any rational number* will return the ball, an infinite number of solutions!

As $m$ and $n$ grow large, the number of collisions grows. Imagine the ball leaving traces on the billiard table. The traces then will literally cover the surface of the table. Now what happens if the slope $m/n$ is an irrational number? Any irrational number can be approximated by decimal fractions, and longer fractions will give more precise approximations. Now, the trajectory with an irrational initial slope is a limiting case of more densely packed trajectories, i.e., large $m/n$. Such an approximate trajectory will then pass very close to some exact trajectory and very close to the initial point. Thus the return (or "almost return") to the initial point is the rule, rather than the exception.

Very dense trajectories that come arbitrarily close to every point on the table, i.e., "everywhere" trajectories, are very important to mathematics and statistical mechanics. They are called *ergodic* trajectories. This billiard table problem is a simple example of Jules-Henri Poincaré's famous theorem stating that any dynamical system will at some time inevitably return very close to its initial state. The theorem is as counterintuitive as this answer must seem to you.

**7. Winter Fun:** A detailed analysis can be quite complex. We need only a "global" approach. What does the duration $T$ of Vanya's ride down depend on? The gravitational acceleration $g$, the friction coefficient $k$, and the geometry of the hill. Because the geometries are similar, the only difference is the height $H$, so that will be, in this case, our geometry parameter. The ride time then must be a function of all these variables, that is

$$T = f(g,k,H)$$

where the functionality $f$ remains to be found.

Now consider the effect of units conversion on length and time variables. In the new units, $T$ will become $T'$ and $H$ will become $H'$. Let the length conversion factor be $a$ and the time conversion factor be $b$, so that $H' = aH$ and $T' = bT$. Because the units of acceleration for $g$ are meters/sec$^2$, the gravitational acceleration $g'$ in the new units has the value $g' = (a/b^2)g$. Since the coefficient of friction $k$ is dimensionless, it needs no conversion factor; it does not depend on the units used. In the new units, the equation becomes

$$bT = f\left(\frac{a}{b^2}\, g,\, k,\, aH\right)$$

The left side of this equation is explicitly independent of the length conversion factor $a$. Therefore, the argument of the function $f$ must include the variables in such a way that dependence on $a$ disappears. This means that $f$ must depend on the ratio $H/g$ but not on the two variables separately. In the "old" variables we must have

$$T = f\left(\frac{H}{g}, k\right)$$

and in the new variables, with converted units,

$$T = \frac{1}{b} f\left(\frac{b^2 H}{g}, k\right)$$

Since the left side of this equation does not depend on $b$, neither can the right side. Therefore the factor $(1/b)$ must be exactly compensated by the function $f$. If this is not so, then the time $T$ to slide down the hill will depend on our choice of units, which cannot be true. Therefore the function $f$ depends on the first variable as the square root of the variable, that is

$$T = \sqrt{\frac{H}{g}}\, K(k)$$

where $K(k)$ is a function of the friction coefficient $k$. Since the snow is the same on both hills, $K$ is really constant in this case.

Now we can tell Vanya that he is correct. If the hill is 4 times higher, the ride down will take only 2 times longer!

The method used to answer Vanya's question is called *dimensional analysis.* As you can see, it is a powerful method for getting answers to complicated questions. If you wish to consider the effect of the weight of the rider and sled, you may also use these methods, with mass included as a variable. You will find that there is no dependence on mass.

**8. Bow and Arrow:** Consider the transfer of elastic energy stored in the stretched bow to the arrow. The stored elastic energy is converted into the kinetic energy of the arrow. The total amount of energy stored in the bow is limited, and therefore the kinetic energy of the arrow cannot exceed this limit. Therefore, the initial velocity of the arrow is also limited. Now consider two related limiting cases, the upper and lower limits for the length of the arrow. An infinitely long arrow will have infinite mass and therefore zero initial velocity, no matter how much energy is transferred from the bow. On the other hand, the arrow can be too short.

The lower limit will be the distance from the bowstring to the center of the undrawn bow. An arrow shorter than this cannot be aimed, and for an arrow *exactly* this length, the bow cannot be drawn, that is, the energy transferred from the bow to the arrow is zero and therefore the intial velocity is zero again.

Thus the distance covered by the arrow is positive, but zero at the two extremes, and there must be, according to the theorem (see "Hints"), at least one maximum. We can visualize how the maximum must occur. If the bow were made of in ideal material, we could bend it double, and this would represent the maximum

elastic energy that could be stored in the bow and transferred to the arrow. The optimum arrow length would then be half the bow length (if straightened) plus half the string length.

We are assuming, for this solution, a simply curved bow. However, many centuries of archery have seen the development of recurved and compound bows, which increase the elastic energy and reduce the optimum arrow size. For such weapons, the choice of materials is rich in complexity. However, the basic principle of this problem applies to all such weapons.

**9. High Technology in Russia:** Yes, it's definitely possible. A circle is not the only figure that has the same thickness in any direction. There are, in fact, an infinite number of shapes with this property. Consider the figure.

The shape joins three arcs of a circle with radius $R$ and three arcs of a circle with radius $r$. The centers of the circles, large and small, are $O_1$, $O_2$ and $O_3$, and the

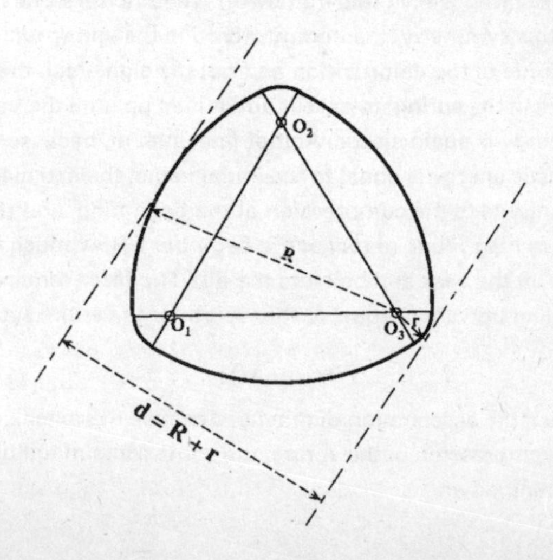

three centers form an equilateral triangle. In *all* directions, the diameter of this shape is $R + r$. Roll a plate on right cylinders with this cross section and the separation of the plate from the floor will always equal $R + r$. This is quite counterintuitive but quite correct. On the other hand, it is fortunate for Ivanov that the boss did not require that the velocity be constant, because the lateral movement of these "rollers" can be quite uneven!

**10. The Jack-in-the-Box (An Exercise in Renormalization):** To begin with, the height to which the Jack and box will jump will be greater if the Jack is depressed with greater force. We will, as recommended in the hint, consider the marginal case where the height is zero, and determine a *minimum force*. For the sake of the analysis, we introduce the spring constant $k$ and the length $x$ of the spring, including the length $x_o$, which is the length in the absence of any applied force. The problem is made difficult by the fact that both elastic and gravitational forces act simultaneously on both bodies. A formal solution (which we will provide) involves some calculations. However, there is another way requiring almost no calculation, that we will present first.

The simplest solution depends on the following property of a linear spring: the effect of a constant force field (such as gravity) on a body connected to the spring is equivalent to changing the length of the undeformed spring. (We will prove this below.) The length change accounts for both the potential energy of the system and the force between the Jack and the box. If we treat the problem by this method, we can ignore (not neglect, but ignore!) the gravitational forces, having already included them. Now the spring properties become *symmetric* with deformation. Extension or compression of the spring will, in this framework, require the same force magnitude $k|\Delta x|$. Because of this symmetry, the energy stored in the spring will now depend only on the magnitude of the deformation and not the sign. Push the Jack down and store elastic energy in the spring; release it and it flies up until the velocity becomes zero and all the energy is again elastic. At that final instant, because energy is conserved and the elastic energy is equal to the initial value, the extension of the spring is identical in magnitude to the compression at the beginning, and *the force on the spring is identical in magnitude to that at the beginning*. How much force at minimum is needed to lift the Jack and box into the air? The force required to lift the entire system by pulling upward is equal to the weight of the entire system, or

$$F = (m+M)g \qquad (1)$$

where $g$ is, as usual, the acceleration of gravity. Because the extension of the spring is identical to the compression of the spring, the same force magnitude will be needed when pushing down.

To prove our initial statement, that the effect of the gravitational force $mg$ boils down *in all respects* to changing the spring length, let an external force $F$ be applied to the Jack, so that the equilibrium length $x$ of the spring is given by the equation

$$mg + k(x - x_o) = F. \qquad (2)$$

Now set $x_1 = x_o - \dfrac{mg}{k}$ and this equation becomes

$$F = k(x - x_1) \qquad (3)$$

which looks exactly as if there is no gravitational field. The potential energy $\dfrac{k(x - x_o)^2}{2} + mgx$ transforms to $\dfrac{k(x - x_1)^2}{2}$ + a constant. Because potential energy is always defined with such a constant, the last term makes no difference to our analysis. This "renormalization" of the length guarantees us all the correct values of the force and a correct form for the potential energy without explicit consideration of the gravitational force. Indeed, gravity disappears from the equations! Renormalization is essential to superconductivity and quantum field theory.

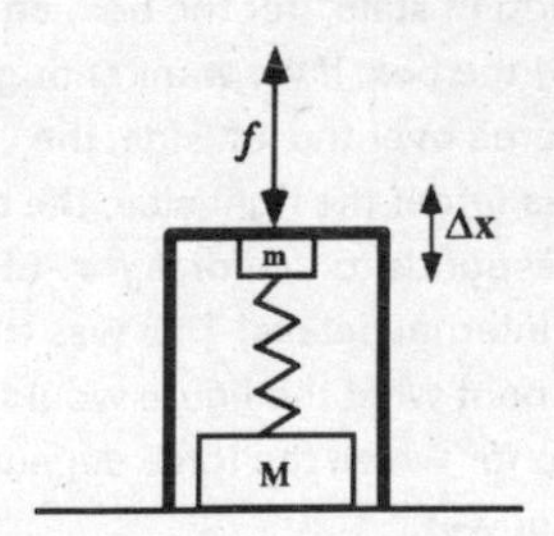


To give you some physical intuition for this procedure, consider the figure. We can try an alternative solution. Imagine the entire Jack-in-the-box assembly to be surrounded by walls; it is a "black box," something thermodynamicists are fond of. We will examine only the properties of the box and proclaim ignorance of the details of what goes on inside the box. What properties will this box have? It will have a mass $M + m$, and will require an upward force of $(M + m)g$ to lift it off the table. If we wish to compress or stretch it in the vertical direction, we will need a force $f = k\Delta x$ to do so, where $\Delta x$ is the change in height of the "box," and $f$ is negative for compression (downward force). By way of example, at liftoff the height of the box has been increased by the amount $\Delta x = (M + m)g/k$, and the force required is $f = (M + m)g$. Knowing only these two properties of the "box," the mass and the stiffness, we now ask how far the box should be compressed to store enough elastic energy to raise it off the table. The answer is shown in the second figure here.

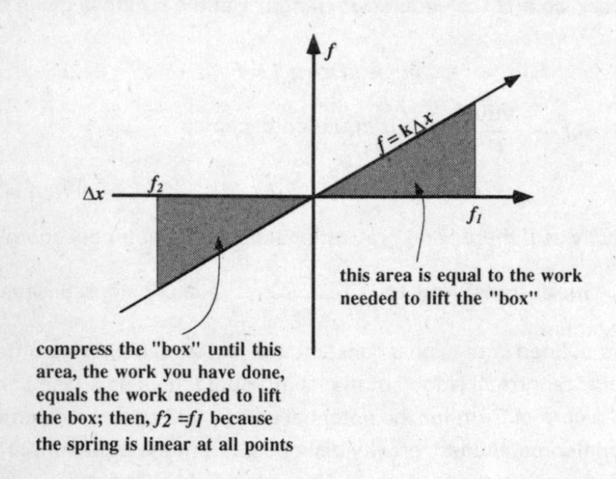


The graph is really the "equation of state" for the box, and the area under the $f = k\Delta x$ line is the work done on (or by) the box. If we want enough energy to extend the box to $\Delta x = (M + m)g/k$, then the area over the left side, the work we did by compression, will have to equal the area under the right side, the energy needed to arrive at $\Delta x = -(M + m)g/k$, which corresponds to the force $f = -(M + m)g$. Please notice that we were able to ignore all the internal details! This was true here because the spring was completely linear. Think about what the figure would look like if it were not! Consider, for instance, a spring for which the force depended on the square of the extension rather than the first power.

And now, for those of you who do not believe in such trickery, here is the formal solution, which is also instructive. We will use three important natural laws. First, Newton's First Law, mechanical equilibrium: when any body is at rest or has constant velocity, the vector sum of the forces on the body must be zero. In addition, we have Hooke's Law: the force necessary to change the length of the spring is proportional to the change in length. At the moment the box is lifted off the table, we can equate the spring force with the gravitational force on the box; by Newton's First Law and by Hooke's Law,

$$k[X - x_o] = Mg \tag{4}$$

It is possible to write down this simple equation because, at that instant, *everything* is motionless; there are no velocities. The maximum length of the spring, which occurs at this instant, is written as $X$. The Jack is at the very top of its trajectory, the spring is fully extended and about to retract, and the velocity of the box is zero; because the box has been fully lifted off the table, there is no reaction force from the table. If this is not the case, we have compressed the Jack too much (in which case it will continue upward and the box will be lifted clear of the table). Also, at the beginning of the experiment, everything is motionless. The spring is compressed to a length $x$ with the applied force $f$ and the weight of the Jack:

$$mg + f = k[x_o - x] \qquad (5)$$

Now there are too many unknown variables (five, in fact), and we need more equations if a unique solution is to be arrived at. These equations will be provided by using a third natural law, the conservation of energy. Since at the beginning and end of our experiment all velocities are zero, the equations for the energies are simple. At the beginning, when the spring is compressed, the total energy is

$$E = mgx + (k/2)[x_o - x]^2 \qquad (6)$$

and at the end, when the spring is fully extended and the box has been barely lifted off the table,

$$E = mgX + (k/2)[X - x_o]^2 \qquad (7)$$

We do not bother to write the gravitational potential of the box, since the box remains on the table or a negligible distance from the table, so its gravitational potential does not change. Now, because the energy in this system is conserved, we can set the energies in equations (6) and (7) as equal and, consequently, with a few minutes of algebra find that

$$2mg = k[2x_o - X - x] \qquad (8)$$

which, with a little bit of work, can be combined with equations (4) and (5) to again obtain

$$f = (M + m)g,$$

a remarkably simple answer, which does not depend on the spring constant $k$.

# Solutions to Chapter 11

## GEOMETRY AND NUMBERS

**1. The Chicken from Minsk versus the Information Superhighway (or Why Did the Chicken Cross the...?)—A Fable of the New Russia:** The farmer pays $6.28. That's right, $6.28! The circumference of the Earth (or any circle) is given by the formula $2\pi R$, where $R$ is the radius. Whatever the radius is, if you increase it by 1 foot, the increase in circumference will be $2\pi$ feet. This is what linear relationships are! Begin with a tiny radius (say, zero) or with a galactic radius measured in parsecs, or *any* radius: increase it by 1 foot and get an increase of $2\pi$ feet in the circumference. (Incidentally, the additional supports cost about $320,000,000. Bureaucrats need geometry lessons.)

**2. Family Fun with Mathematicians:** First, the problem is meaningful *only* if all the ages are integers. You *must* assume that! The next step is to go ahead and use the first condition, that the product of the sons' ages is 36. This gives us just 8 options:

| age 1 | age 2 | age 3 | sum |
|---|---|---|---|
| 1 | 1 | 36 | 38 |
| 1 | 2 | 18 | 21 |
| 1 | 3 | 12 | 16 |
| 1 | 4 | 9 | 14 |
| 1 | 6 | 6 | 13 |
| 2 | 2 | 9 | 13 |
| 2 | 3 | 6 | 11 |
| 3 | 3 | 4 | 10 |

Among these options is the correct solution. We must choose. Calculations of the sum (the fourth column) show the only possible numbers of windows in the house. If the sum were 38, 21, 16, 14, 11, or 10, Igor would have been able to solve the problem immediately. He was unable to do so *only* because the number of windows in the house (and the sum of the ages) was 13! Because of this, he did not have a unique solution until Pavel informed him of the hair color of his "older son." If the ages are 1, 6, and 6, there are older *sons* but not an older *son,* and the ambiguity is resolved: Pavel's sons are 2, 2, and 9 years old.

This problem is important because it shows many of the major steps in solving any problem:

(a) Pay attention! Every statement, every fact, may have significance. For example, we may ask, "why mathematicians?" This is an important point because we suspect that an answer can indeed be found, but if the mathematician Igor says that it is impossible, then the answer either does not exist *or* the solution is not unique. This is the key to the solution!

(b) You may have to make certain assumptions in order to find any answer. In this case, the assumption is that the ages are integers (or can be rounded off, as people often do). The fact that mathematicians are involved is an important hint in this direction!

(c) The fact that you do not know how to solve the problem must not discourage you. This is a normal condition unless the problem is trivial.

(d) You should immediately solve the part of the problem you understand, even if it is only a small step. It may even be formalization, restatement, or reformulation of one of the conditions of the problem. Even if you cannot solve the whole thing, *start to solve the problem!*

(e) When your solution has advanced, you should look for the hidden meanings of the data, for instance, the son with red hair. The common words and phrases should be unscrambled and their hidden meanings deciphered. This is the rule rather than the exception in actual problems as well as teaching problems.

(f) We solved this problem from both ends rather than "the beginning." This is a very helpful technique in problem solving. On the way, we had an experience familiar to many scientists, that is, we learned something new and unexpected: there are exactly 13 windows in the building.

**3. Some Numbers!:** The prime factors in 24 are 3 and 2, that is, $24 = 3\times2^3$. Thus we need to prove that $(n^2-1)$ is divisible by 3 and 8, or 6 and 4. Let us try the first pair. If we observe that $(n^2-1) = (n-1)(n+1)$, we can proceed to the most important step in the solution. The numbers $(n-1)$, $n$, and $(n+1)$ are consecutive natural numbers,* e.g., 5, 6, and 7. In any three consecutive natural numbers, one of the three numbers must be divisible by 3. Since n is a prime number, one of the other two numbers is divisible by 3.

Now to prove that $(n^2-1)$ is also divisible by 8. First, note that $(n+1)$ and $(n-1)$ are both even numbers, since $n$ is a prime number greater than 3 and therefore is an odd number. We can now show that one of these two numbers is divisible by 4. The two numbers $(n+1)$ and $(n-1)$ are consecutive, even natural numbers, and of any two consecutive, even natural numbers, one must be divisible by 4. Therefore the product $(n+1)(n-1)$ must be divisible by 8. Since $(n+1)(n-1)$ is also divisible by 3, it is divisible by 24.

**4. The Root of All Our Problems:** Two. A variation of the last number, that is, the numeral in the innermost square root, must have a small effect on the value of $x_n$. The greater the number of roots $n$, the smaller the effect of the variation. Now consider an intuitive solution: if the innermost number is replaced by a 4, the whole expression unravels and the answer is $x_n = 2$. If $n$ is large, the replacement should have a small effect.

If this is unsatisfying and you want a formal proof, we offer this in two steps. First, we must prove that there is a limit for $x_n$ as $n$ approaches infinity.We can prove the existence of the limit by using the theorem that a monotonic sequence always has a limit if it is bounded. (The intuitive solution proves that $x_n<2$.) We can also easily see that $x_n$ is monotonic and increasing, that is, that $x_n$ increases with $n$. To see this, simply add $\sqrt{2}$ to the innermost root. Now that the existence of a limit has been proven, we can go to the second step, which is to determine the limit. We rewrite the equation for $x_n$ as

$$x_n = \sqrt{2 + x_{n-1}}$$

and observe that in the limit, as n approaches infinity, both $x_n$ and $x_{n-1}$ tend to the same limit, say, $x$, so $x_{n-1} \rightarrow \infty$ and $x_n \rightarrow \infty$ and this equation reduce to

$$x = \sqrt{2 + x}$$

which has only one positive root, $x = 2$. (Square both sides and solve the resulting quadratic equation. The roots are -1 and 2.)

**5. Scaling the Pythagorean Theorem:** Consider the figure on the next page. The right triangle $ABC$ has been divided by the altitude $CD$ into two right triangles $ACD$ and

*A natural number is a positive integer, zero not included.

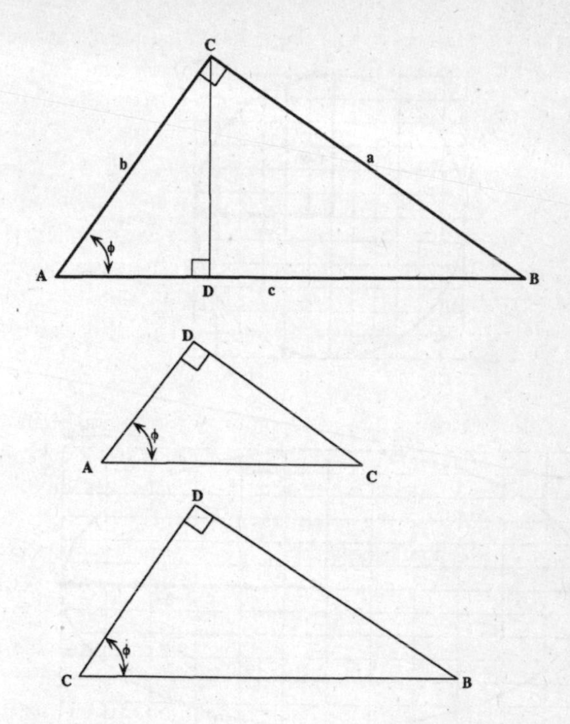


$CBD$, and all three triangles are similar. All three are right triangles and have in addition the same angle $\phi$ at one vertex. Now we need to use only the fact that areas of similar triangles are proportional to the squares of similar sides. Each side of the original triangle $ABC$ is a hypotenuse of a similar triangle. Therefore the areas of triangles $ABC$, $ACD$, and $BCD$ can be represented as $Kc^2$, $Kb^2$, and $Ka^2$, respectively, where $K$ is a constant. Since the area of $ABC$ is the sum of the areas of $CBD$ and $ACD$, that is, area ($ABC$) = area ($CBD$) + area ($ACD$), we have

$$Kc^2 = Ka^2 + Kb^2.$$

Canceling out the constant $K$, we have the Pythagorean Theorem:

$$c^2 = a^2 + b^2.$$

**6. Stretching Olya's Mind (or A Stretch of Olya's Imagination):** Begin with an educated guess. Consider a circle as a special case of the ellipse where the semi-axes $a$ and $b$ are equal ($a = b = r$). The area of a circle $S = \pi r^2$ or $S = \pi r \cdot r$ is then just a special case of the formula $S = \pi ab$. This appears to be a reasonable expression because $a$ and $b$ enter this equation symmetrically.

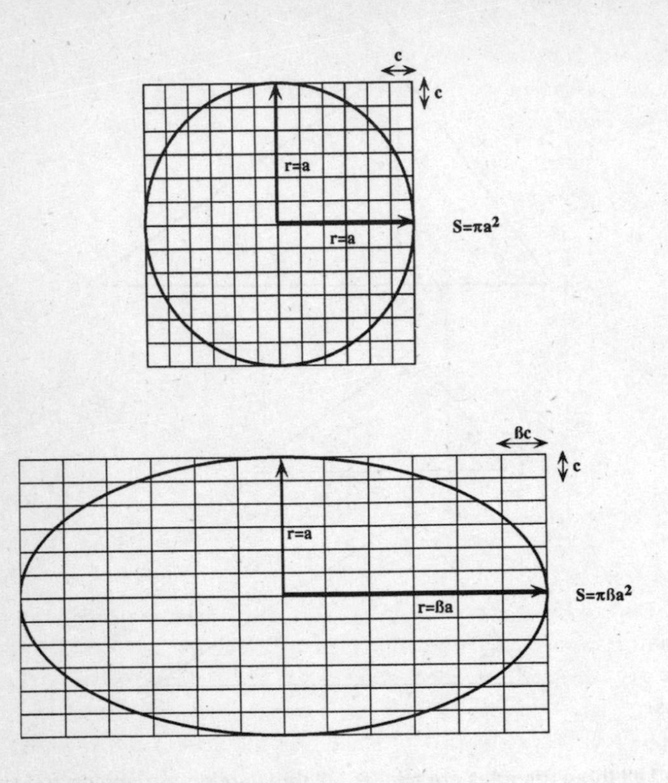


However, Olya did more than guess. She observed that an ellipse can be obtained from a circle by uniform stretching in a single direction. Now consider uniform stretching of a square with side $c$, along one of the sides. Before stretching, the area of the square is just $c^2$. After stretching uniformly along one of the sides (say, the base), the length of the side will become $d$ and the area of what is now a rectangle will be $cd$. The procedure can be described as stretching by a factor $\beta = d/c$, with the surface area increasing from $c^2$ to $\beta c^2$. Olya realized, as she stared at the circle drawn on her drafting pad with the grid of small squares, that any geometrical figure can be drawn on such paper and divided into small squares. If now the paper is uniformly stretched along one of the sides, each square will become a rectangle and its area increased by the factor $\beta$. Consequently, the area of the entire figure will be increased by the factor $\beta$. If this figure were a circle, its area would increase from $\pi a^2$ to $\pi \beta a^2$ or $\pi ab$, since $b = \beta a$. The figure illustrates what Olya saw.

**7. The Problem That Didn't Fool Von Neumann:** Look at the graph, which plots coordinates versus time for the bicyclists and the dog, and shows the distances AB between the bicyclists.

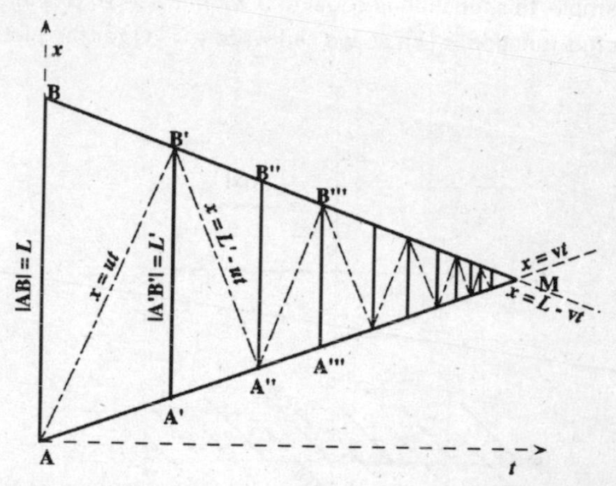


The triangles $ABM$, $A'B'M$, $A''B''M$, etc., are all similar triangles. We will call the similarity coefficient or scaling factor $q$. When the bicyclists are distance $L$ apart, the dog will need a running time of $L/(u + \mathrm{v})$. During this time, the bicyclists come closer together, closer by $2\mathrm{v}L/(u + \mathrm{v})$, so the new distance between them is $L' = L - 2\mathrm{v}L(u + \mathrm{v}) = L(u - \mathrm{v})/(u + \mathrm{v})$. Therefore, the scale factor we need is $(u - \mathrm{v})/(u + \mathrm{v}) = q$. The dog's second run will be of length $qL/(u + \mathrm{v})$. We have to repeat this situation over and over again—in fact, we have an infinite series for the total length of the dog's run:

$$\textbf{distance} = \left(\frac{Lu}{(u + \mathrm{v})}\right)(1 + q + q^2 + q^3 + \ldots)$$

The series is readily summed up:

$$\textbf{Sum} = 1 + q + q^2 + q^3 + \ldots = \frac{1}{1 - q} = \frac{u + \mathrm{v}}{2\mathrm{v}}$$

Combining these two equations with the value we obtained for $q$, we obtain the answer: the distance covered is $Lu/2\mathrm{v}$, the same answer as in problem #10, chapter 2. The situation does not change very much if the velocities of the bicyclists are different. In view of the symmetry and simplicity of this solution, it appears to us that Von Neumann must have seen it (or possibly a better one!). We have therefore renamed this problem "The Problem That Did Not Fool John Von Neumann." (We would like to thank David Hicks and the faculty of the Saint Paul's School for bringing the original anecdote to our attention.)

**8. The Party Secretary's Fence:** Here is the most elementary and interesting solution: use the function called the "integer part of $x$," which is equal to the greatest integer, which is equal to or less than $x$. Some call this function "round towards minus infinity." For example, this function is equal to 5 when $5 \leq x < 6$, 6 when $6 \leq x < 6$, etc. The notation for this function is $[x]$, so we can write $y = [x]$ for the plot shown in part (a) of the figure.

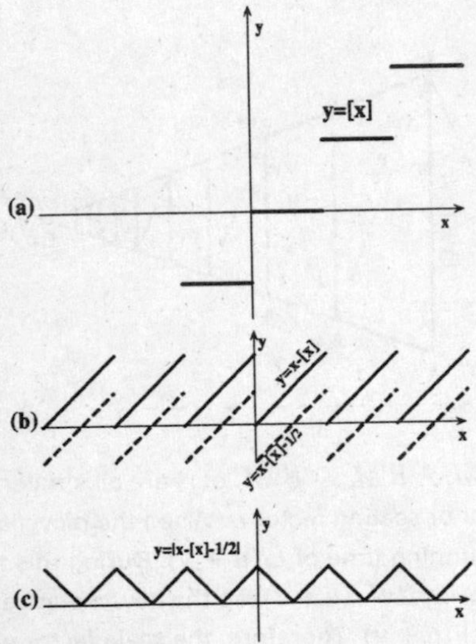


We need very little more, now. The following steps will give us the shape for the top of the fence:

(a) subtract $[x]$ from $x$ to obtain the function $y = x\text{-}[x]$, which is shown by solid lines in part (b) of the figure;

(b) Subtract 1/2 from $y = x\text{-}[x]$, to obtain $y = x\text{-}[x],\text{-}1/2$, shown by dashed lines in part (b) of the figure;

(c) finally, take the absolute value of $y = x\text{-}[x],\text{-}1/2$, which immediately gives us the sawtooth shape we need, as seen in part (c) of the figure.

Scaling factors can be introduced to make the height of the sawteeth conform to the Party Secretary's expectations and to increase or decrease the spacing.

If you are more analytically minded, you might like the following solution better, because it is based on more familiar periodic functions. The inverse sine or arcsin function is defined by the condition that for $-\pi/2 \leq x \leq \pi/2$, the following identity holds:

$$\arcsin(\sin x) \equiv x.$$

In fact, the equation $y = \arcsin(\sin x)$ describes the fence very well for $-\pi/2 \leq x \leq \pi/2$. The symmetry of the sine curve about $x = \pi/2$ (or trigonometric identities, e.g., $\sin(\pi/2+z) = \cos z = \sin(\pi/2-z)$, where $z$ is any variable) makes the curve reverse at $x = \pm\pi/2$ and gives us the sawtooth shape, out to $x = 3\pi/2$. Finally, the periodicity of the sine function makes the whole thing repeat for the entire real $x$-axis.

On the other hand, if you are a very modern engineer (the kind who uses a computer to generate a fast Fourier analysis before adding the quantities $a$ and $b$, using the same computer to deconvolute the results), you have probably thought first of the Fourier series solution. Yes, you can do it that way. If you have thought of this method, we will leave the details to you, for you have certainly earned the right to work that hard. For instance, a solution that is symmetric with respect to the origin and has peaks at multiples of $\pi$ is

$$y = \frac{4}{\pi} \sum_{k=0}^{\infty} (-1)^k \frac{\sin(2k+1)x}{(2k+1)^2}$$

The masochism involved in seeking a solution this way would seem to be mitigated by the flexibility of the method, which can represent a fence of any shape (provided the Party Secretary's goat doesn't eat your computer). However, by using either of the first two solutions you can also do this. For instance, take for the first solution $y = g(|x-[x]-1/2|)$, where $g(x)$ defines the shape of an individual tooth or a portion of the tooth; or, for the second solution, let $y = g(\arcsin(\sin x))$ to give you whatever shape you introduce with $g(x)$, suitably repeated. And if you are dealing with a fence of finite length, rather than the Party Secretary's grand structure, you will find that boundary conditions are a real nuisance with Fourier series. They are much easier to live with in the case of the first two solutions. In solid-state physics, where finite periodic structures (known as crystals) are the rule, it is often necessary to introduce special boundary conditions ("periodic" boundary conditions) in order to live with Fourier series or transforms.

This problem is dedicated to a friend, Rein Tomsaluu of Talinn, a highly regarded applied mathematician who often labored summers as a highly paid (relative to mathematicians) carpenter. In the fall, he went back to science. He enjoyed both greatly.

**9. What Color Was That Bear? (A Lesson in Non-Euclidean Geometry):** This is a generalization of a familiar problem, which goes like this: the camper leaves his tent and does exactly what our camper did, walking 2 km South, 5 km West, and 2 km North, and finds himself back at his tent. There he sees a bear eating his dinner. What color is the bear? The answer is "white," because only at the North Pole, in the Arctic,

would these three routes form a closed trajectory, and the bear must be a polar bear.

Now you must ask whether the final location of our camper depends on where he starts. The answer is yes, definitely. To obtain the answer to our question, we will consider the worst- and best-case scenarios (worst and best from the point of view of the camper: closer to the tent is closer to dinner!). Now we are confronted with a deep and very interesting question: What do we mean by distance on the surface of a sphere? Because we are dealing with a spherical surface, the geometry is not Euclidean. The final answer is an inequality: $X_{min} \leq x \leq X_{max.}$ The problem now boils down to finding the limits, the minimum walking distance $X_{min}$, and the maximum walking distance $X_{max.}$

**The best case (minimum distance)**

This case corresponds to the riddle about the bear: if the camper's tent is exactly at the North Pole, he will arrive at his tent after the final leg of his hike. This is easy to see because of the spherical shape of the Earth, and the way we measure latitude and longitude. In this case, as shown in this figure, the distance in question is zero. We shall state this formally as $X_{min} = 0$. (See figure 1.)

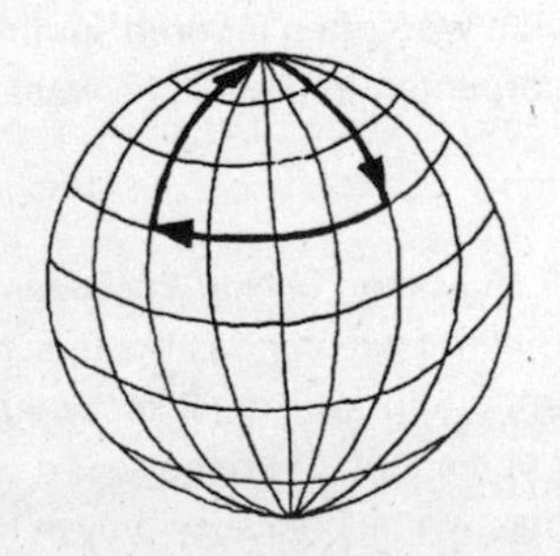

FIGURE 1

This result is drastically different from what one would expect from our usual experience or just from Euclidean geometry. The most Euclidean case arises when the camper starts 1 km North of the equator (the camper would cross it when he or she is halfway South). In this case, the camper's path consists of three consecutive sides of what is practically a rectangle, so the distance to the tent would be exactly 5 km (with no prospect of an early dinner in this case either). But going farther South makes the situation even worse. There we are indeed approaching the worst-case scenario.

**The worst case (finding $X_{max}$)**

Now pitch our camper's tent somewhere near the South Pole. The camper walks southward 2 km and then westward for 5 km. However, now the camper's latitude changes dramatically, in this case by 180°, as illustrated in the next figure.

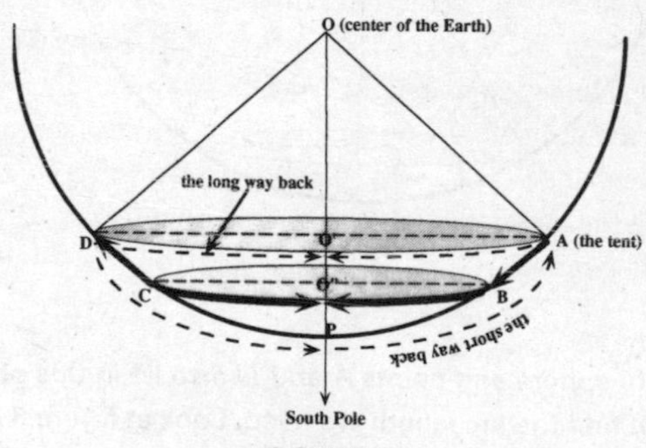


FIGURE 2

The camper begins at point $A$, proceeds southward to $B$, and then westward to $C$, and north to $D$, *the same distances as before.* Now if we have placed the tent so that the westward leg does change the latitude by 180°, point $D$, rather than being back at the tent (point $A$), is the farthest of all points from $A$! The shortest way back to the tent from point $D$ turns out to be the arc $DCPBA$, a great circle that passes over the South Pole. In fact, you may remember that *the shortest path between any two points on the surface of a sphere is the arc of the great circle connecting these points.* This is the way distance is defined in the general case of non-Euclidean geometry. This path $DCPBA$ is almost a straight line because the radius of the sphere is the Earth's radius (6378 *km*). If the camper would go East from $D$ to $A$ along the parallel, it would be much longer. This path would also be a semicircle, and the optimal path $DCPBA$ is very close to the diameter of the semicircle. Therefore, the optimal path is about $\pi/2$ times shorter. We will go into some detail now, so you may wish to go to "more best cases" and the ensuing discussion.

We now need to find the length of the arc $DCPBA$. We know the lengths of arcs $|AB| = S$ and $|CD| = N$ (2 km each), so we need only to find the *length of the arc* $|CPB| = W$. *Now, the westward leg of the camper's hike is* $BMC$, *an arc of a lesser circle* (half of a parallel on the globe), where the circle lies in a plane perpendicular to the axis of the Earth. The radius $O''B$ of this circle can be easily found when we realize that the arc $BC$ is a semicircle with length W km (the camper's westward leg). Therefore, the radius of the semicircle is $W/\pi$.

Now we can find the arc $BPC$: we cut the sphere with a vertical plane passing through the Earth's center O and the points $B$ and $C$, as shown in figure 3.

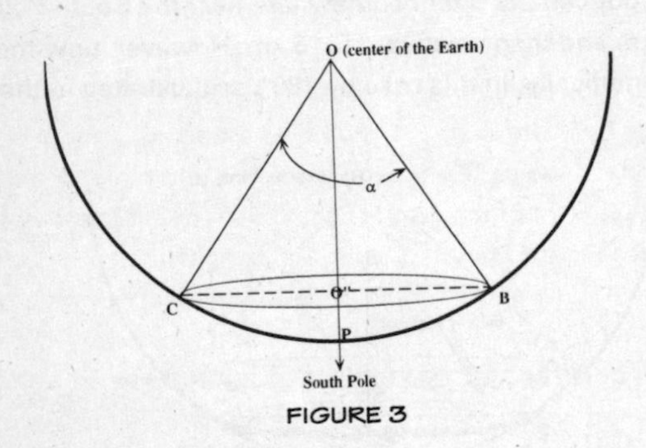


FIGURE 3

The axis of the (Earth) sphere and points $A$ and $D$ also lie in this plane, as shown in figure 2. Now we can find the arc length we need. Look at figure 3 again: $OB$ and $OC$ (and $OA$ and $OD$, for that matter) are all radii of the sphere, and the angle between $OC$ and $OB$ can be calculated by plane trigonometry since the distance $BO''C$ is known: twice the semicircle radius, or $2W/\pi$ kilometers. Therefore, as shown in figure 3, the angle $\alpha$ between $OC$ and $OB$ is given by

$$\alpha = 2 \arcsin\,[W/\pi R]$$

where $R$ is the radius of the Earth. Now the arc $BPC$ is simply $\alpha R$, as it is an arc of a great circle of radius $R$ with the subtended angle $\alpha$. Because $A$ and $D$ also lie in this great circle (look at figure 2), the arc $DCPBA$, the shortest way back to the tent, is just $\alpha R + |AB| + |CD|$, or $\alpha R + 2S$ km. Using the above expression for $\alpha$, we obtain $X_{max,}$ which determines the worst case, in the form

$$X_{max} = 2[S + R \arcsin(W/\pi R)].$$

As the radius of the Earth is approximately 6,400 km, we can simplify this worst case with high precision. Because the Earth's radius *R* is larger by far than any distance the camper is likely to walk, the argument of the arc sine, $W/\pi R$, is quite small. For small $z$, $\sin z \approx z$ and arc $\sin z \approx z$, in this case to a high degree of precision. Then our formula simplifies to

$$X_{max} = 2(S + W/\pi).$$

Now it is easy to estimate the numbers. Taking $S = 2\text{km}$ and $W = 5\text{km}$, as the story tells us, we find $X_{max} = 7.183\text{km}$. As the camper has already walked 9 km on the three outbound legs, dinner has now receded into the distance!

**And now, some more best cases**

We have purposely chosen the worst case so that the westward leg changes the longitude by 180°, as we knew that this would place the camper farthest away and cause the maximum inconvenience (this principle is not unique to Soviet governance, but seems to be a universal property of bureaucracies). Now reconsider figure 2: hold $S$, $N$, and $W$ constant, but move point $A$ closer to the South Pole. Arc $BC$ now changes the longitude by more than 180° and the camper will end up closer to his or her tent at point $A$. In fact, we can choose to place the tent close enough to the South Pole that arc $BC$ is 360° and points $B$ and $C$ coincide. The northern leg is now the reverse of the southern leg and points $D$ and $A$ also coincide. The camper ends up back at the tent, and we have another best case (but the camper may now find a penguin in his tent)! We can continue to move the tent southward until arc $BC$ is 720°. We will obtain a best case whenever arc $BC$ is a multiple of 360°.

How close need we be to the South Pole for these solutions to occur? Reexamination of figure 3 leads to the conclusion that

$$W = 2\pi n|O''C| = 2\pi nR\sin(\alpha/2)$$

where $n$ is any natural number (a non-negative integer). Then the angle $\alpha/2$, which is the latitude difference from the South Pole to point B or C, is

$$\frac{\alpha}{2} = \arcsin\left(\frac{W}{2\pi nR}\right)$$

and because $R$ is so much larger than $W$, we can again simplify, this time to $\frac{\alpha}{2} = \frac{W}{2\pi nR}$ so the distance of the southmost point of the path from the South Pole is $|PB| = R\frac{\alpha}{2} = \frac{W}{2\pi n}$ In our case, $W$ is 5 km and our camper first gets lucky when $n = 1$, when point $B$ is 0.7958 km = 795.8m (which is about 1/2 mile) from the South Pole. The tent must be pitched about 2.796 km from the South Pole; for $n = 2$, the tent must be 2.398 km; for $n = 3$, 2.265 km, and so on. There is an infinite series of such optimistic best solutions as we approach the critical distance of 2 km from the pole.

The most interesting thing in this sequence of numbers is that setting the tent in a point somewhere between these lucky points would result in a distance to walk that is similar to that of the worst-case scenario. For instance, siting the tent exactly between any two "lucky" numbers would put the camper 180° away, on the other side of the Earth. As the number $n$ increases, the distance between these lucky and unlucky settings would become less and less. So a small misstep in setting up the tent may completely change the probability of a timely dinner. When the initial setting position is just a little more than $S$ km from the South Pole, then even a minor error in the initial position of the camper may result in a considerable change in the camper's position at the end of his walk. The latter actually becomes completely uncertain. Such a situation is called "sensitivity to initial conditions," and is one of two requirements for chaos, which is a very popular topic at this time. (The other requirement, which leads to fractal dimensions, leads to very chaotic discussions!)

**Making the best of the worst case**

Now return to the worst case. The camper has gone South, West, and North, but does not know his or her position precisely. We have seen that it may well happen because of the sensitivity to the initial conditions. It is getting dark and cold. Supplies are running low. How does the camper avoid missing the tent altogether? What is the best strategy that guarantees that the camper won't miss the tent? The best strategy would be to go directly to the East. This path is shown in figure 2 as "the long way back." We can calculate the length of this arc, using the same approximation as above: the length of the arc is $\pi/2$ times the diameter $DO'A$. However, on this scale, the Earth is practically flat and therefore $DO'A \approx X_{max.}$ Thus the surest way to return is only longer than the shortest way by a factor of $\pi/2 \approx 1.57$.

**Now a couple of comments to conclude this long story**

The most famous Russian military leader was Aleksandr Suvorov, a general in the late eighteenth century. He was particularly renowned for fast, long marches of his troops, and would formulate his military doctrine into riddles, such as "Which road is the shortest?" His answer to this riddle was: "The highway always is." The underlying message is that the highway is the surest way to your destination that, in the final analysis, is most often the shortest way in terms of time.

The kilometer is an excellent measure of distances on the surface of the Earth: 1 km is defined as $10^{-7}$ of the half meridian. This distance was actually measured by a group from the French Academy of Sciences in the early nineteenth century (approximately 1810), by traveling a few angular degrees across the Sahara Desert and measuring the distance traveled with an odometerlike device. (One of the group, E. Malus, a very talented physicist, contracted a fever during this trip and died soon after his return to Paris.)

We have conveniently ignored in this problem the fact that the Earth is not perfectly spherical. Deviations from sphericity have been measured with great precision by cartographers and are of great concern to navigators and astronauts, among others.

**10. Identity Crisis:** The South magnetic pole. Of the two poles of a magnet, one was by convention called the North Pole. If used as a compass needle, the North Pole points North and there is no alternative to naming the poles of the magnet. However, it is the opposite poles of magnets that attract. Similar poles (North-North, for instance) repel. Therefore, the magnetic pole of the Earth that attracts the North Pole of the compass needle is the South magnetic pole. We can see then that the South magnetic pole must be near the North geographic pole and the North magnetic pole must be near the South geographic pole! As these conventions have been in use for centuries, straightening this situation out is a hopeless task!